LOBSTERING
off
CAPE ANN

LOBSTERING off CAPE ANN

A Lifetime Lobsterman Remembers

Peter K. Prybot

Charleston · London
History PRESS

Published by The History Press
Charleston, SC 29403
www.historypress.net

Copyright © 2006 by Peter K. Prybot
All rights reserved

Cover Image: Rainbows are often the finale to summertime squalls, as shown by this Pigeon Cove Harbor shot.
All images courtesy of author unless otherwise noted.

First published 2006

Manufactured in the United Kingdom

ISBN 1.59629.085.4

Library of Congress Cataloging-in-Publication Data

Prybot, Peter K.
 Lobstering off Cape Ann : a lifetime lobsterman remembers / Peter K. Prybot.
 p. cm.
 Includes bibliographical references.
 ISBN 1-59629-085-4 (alk. paper)
 1. Lobster fisheries--Atlantic Coast (North America) 2. Lobster fishers--Atlantic Coast (North America) 3. Prybot, Peter K. I. Title.
 SH380.2.A75 2005
 639'.540974--dc22
              2006003967

Notice: The information in this book is true and complete to the best of our knowledge. It is offered without guarantee on the part of the author or The History Press. The author and The History Press disclaim all liability in connection with the use of this book.

All rights reserved. No part of this book may be reproduced or transmitted in any form whatsoever without prior written permission from the publisher except in the case of brief quotations embodied in critical articles and reviews.

Contents

Acknowledgements	7
Introduction	9
I. The Workforce	11
II. An Inshore Lobsterman's Year in Review	25
III. Rare Personal Experiences	40
IV. Lobster Tidbits	49
V. The Harvesting of Lobsters and the Big Changes	69
VI. Lobster Marketing	108
VII. The Lobster Regulations	128
VIII. Lobster, the King of Cuisine	139
Conclusion	146
Bibliography	159
About the Author	160

Acknowledgements

This book is dedicated to a number of important people. First and foremost, to my wife Anne and son Tom for their tolerance during this two-year book writing, editing and re-editing process. This book would never have come about if it weren't for my wife's keyboarding, computer and editing skills. Also, thanks to my late parents Roman J. and Eleanor Prybot for the gifts of life, creativity, curiosity, perseverance and appreciation, and my brother John for love of nature. I also thank all of the lobstermen and scientists who believed in me and gave me their time and information for this book. Lastly, this book is dedicated to all my deceased lobstermen friends whose spirits are still out there on the fishing grounds.

Introduction

Like the trademark "Three Cs—clear, calm and cold" of an overhead high pressure system during a winter's night, the American lobster also has three trademark Cs among its harvesters, purveyors, consumers and fair portions of the world's people. It is the king of cash, cuisine and curiosity.

The American lobster continues to be the most valuable commercially important species for the American northeast coast and Atlantic Canada. Recent National Marine Fisheries Service statistics list annual U.S. lobster landings at around ninety million pounds worth approximately $366 million. Canada's Department of Fisheries and Oceans recorded their recent lobster landings and values as fifty-three metric tons worth $645 million Canadian.

Lobster consumers now span the globe. Thanks to globalization, the lobster has become part of more peoples' cultures today. Many people simply love the cold-water lobster's taste and meat texture and also the fun of getting the meat out of the lobster. Amateur and professional chefs not only continue to prepare and serve the shellfish traditionally, but also create new lobster dishes. New markets wait to be opened.

Lobster is the king of curiosity for those not only on the inside but also those on the outside of the industry. The inquisitive public poke their noses right up against the glass of lobster tanks and are smitten by these creatures, especially odd-colored or huge ones. The public is also curious about the harvesters, the fishing gear and what goes on out on the ocean. Along the waterfront, visitors often query lobstermen: How many lobsters do you catch in a trap? Have you ever gotten bitten? And how can you tell a female lobster from a male? How many lobsters did you catch today? My favorite answer to that last question is, "I've done better on some days and worse on other days."

In addition, lobstermen continually try to understand and think like their quarry to better figure out the lobster's whereabouts, movements and habits to further outsmart them with better traps and baits.

"The business has changed, but not the lobster," say today's lobstermen. The once slow-paced lobster fishery has recently largely done a 180-degree turn for its harvesters and handlers, thanks in part to technological advances, more leisure time and globalization. The Atlantic coast, like the world, has gotten smaller, with commercial fishermen now finding themselves sharing more of the coast and the ocean and its common property resources with growing numbers of recreational boaters and fishermen, and new coastal residents. The once independent lobsterman has become more of a team player, technocrat and eco-warrior. Even though the lobsterman might not necessarily be the cause of a stock decline, he's constantly the target of regulators and environmentalists and has had to adapt to new rules.

As bad as this might be at times, the lobsterman still has most of his freedoms of the ocean and fishing, unlike his scallop and ground fishermen peers, who have limited days at sea to fish and have to call in and out for every trip. Many of these fishermen also have to equip their vessels with visual monitoring systems so that the government can follow all of their moves at sea.

Technology has made the lobsterman's job easier, and his job is in great demand today. But lobstering, like all fishing, is a game of persistence. Those who stick out its ebbs and flows will make out all right.

Ironically, U.S. lobster stocks have been classified as over-fished by many regulators and biologists; yet Maine has had recent record annual landings in the sixty to seventy million pound range. The numbers might have been even higher a few years ago thanks in part to more lax reporting requirements. Inshore lobsters from Cape Cod south down to Long Island Sound have experienced setbacks.

Lobstering off Cape Ann: A Lifetime Lobsterman Remembers not only records historical changes in the industry and puts the spotlight on some of the pioneers who have led the way, but also gives the outsider the inside story of the lobster, its harvesters, handlers and the industry in general and what's in store for them. Although this book has been in the making for years by recording and photographing specific events, a flurry of other lobster book publications have held off its printing.

The year 2006 marks the start of my forty-sixth lobster season. Where has the time gone? I still vividly remember my first time out lobstering, struggling to haul that first trap over the rail and pegging the claws of my first legal-sized lobster. Although lobstering has largely consumed my life since age twelve and nearly taken it at least once, it's also given me a livelihood, a hands-on education, a defined and meaningful purpose in life, a unique sense of freedom and pride and a job that I wouldn't trade for the world. I'm one of the few that can say that I love what I'm doing. This is where the lobster and the industry stand as of 2006.

I
The Workforce

Every lobsterman has his or her story of how and why they are in business. Some were born into it, and the lobstering is just in them, often generations worth on both sides of their family, while others were outsiders who gradually worked their way into the business and up in it, too. Nova Scotian lobsterman Jerry Himmelman and Deer Isle, Maine, lobsterman Steve Robbins III each represent five generations of lobstermen. Richard Nickerson takes the cake for having one of the longest fishing lineages. "My ancestors came over on the *Mayflower*, and they made their way over to Nova Scotia in 1750," he says. Nickerson is at least the ninth to possibly tenth generation fisherman in his family. He lobsters in Lobster Fishing Area 34 off Nova Scotia. These men are examples of how lobster ports are frequently dominated by certain long-time fishing families.

"Beal, Smith, and Alley are the big names at Jonesport/Beals. Beal is probably number one," says Randy Beal of Jonesport, Maine, who lobstered and dragged for quahogs there and now fishes out of Cape Ann, Massachusetts. Jonesport and the neighboring township of Beals constitute a big lobster fishing area in down east Maine. Jonesport/Beals also has some of the finest boat builders and designers in America.

Collectively, the main lobstering workforce, which is located from the United States mid-Atlantic coast to the Maritimes, numbers in the thousands, is multi-ethnic, runs the gamut of personalities, represents all levels of education, and is getting older—probably averaging in the late forties thanks to greater limited entry into the lobster fisheries. Most will stay in the business until they are either no longer able to move about, or they die. The majority have thought about fishing since they were kids. They climbed their fishing ladders by first fantasizing about fishing and constructing make-believe boats out of driftwood and Styrofoam and playing with these at the water's edge, and then going out on others' boats before going out on their own. Novices learn from trial and error and from their peers. Families pass on the tricks of the trade to their children. Everyone improves with time.

Lobster harvesters possess strong hunter-gatherer instincts and love of self-employment, physical outdoor work and freedom of the job and ocean, along with its moods. We also keep hoping for that big catch and love the challenges of making it on our own, working close to nature, and even trying to fully understand the lobster and outsmart it. We like making good money, too. We also like to talk the universal language of fish with other lobstermen and learn about their gear and fishing. We are proud of our coves and harbors. When we introduce ourselves to other fishermen, we say right away, "I fish out of such-and-such a harbor." Every cove and harbor has a history and is known for certain things. Gloucester Harbor is known the world over as America's oldest fishing port. Our main job is to hunt down the lobsters, bring them up from the bottom, and enter them in good shape into the market.

We lobstermen are competitors. We don't like to share our grounds and lobsters with others and often enjoy seeing another man lose his traps (but only when we don't lose ours). But we'll come to the rescue of our peers when they are in need, and give them the shirts off our backs.

Colby Polisson of Gloucester, Massachusetts, born in 1993, represents the low end in the age range of today's lobstermen. He is one of those natural-born fishermen who already exhibits many of the lobsterman traits. Colby began helping his grandfather aboard his forty-two-foot dragger-lobster boat *Rhumboogie* out of Pigeon Cove Harbor, Massachusetts, at age seven. On fishing days, Colby gets up before dawn during the summers and school vacations. Out fishing, he fills bait bags, empties traps and bands lobsters. "He will reach right into the trap. Sometimes, his hand will come out with lobsters clinging onto his gloved fingers," chuckles his grandfather, Jason Polisson, also of Gloucester. Colby doesn't get seasick. He also helps his grandfather work on traps ashore. Colby already has a small skiff and ten traps of his own. "I would like to be a lobsterman," he says.

During the off-season for lobstering, Colby goes out dragging the whole day, even in the winter. His grandfather alternately nets northern shrimp and groundfish such as cod, haddock and flounder. Groundfish live most of their lives on or near the bottom of the ocean.

In 1999 I had the privilege of meeting Maine's longest license holder, ninety-six-year-old Millard Crowley of Beals, Maine, who has held his license for eighty-six years. He represents the high end of the lobstermen's age range. Crowley started lobstering rowing a "pea pod" and hand-hauling approximately seventy-five single wooden traps marked by his black and white buoy colors. Crowley built his last boat himself, the thirty-foot *Aura V*.

"Lobstering is what keeps him going," says his son Vertland, also a lobsterman. Millard still lobstered until age ninety-seven. On December 12, 2002, Crowley celebrated his one-hundredth birthday. Sad to say, he has since passed away.

I'm one of those lobstermen who gradually worked his way into lobstering despite having neither any fishing family history nor roots in Gloucester. I was a true outsider who simply fell in love with lobstering as a twelve-year-old back in 1960. I probably inherited my father's love of recreational fishing and was also influenced by my surroundings.

The Workforce

This drawing by Joe Sinagra of Gloucester depicts how I used to transport a trap from my house to Lane's Cove. The trap was tied to one side of the bicycle, and I balanced myself on the other side during the mile coast downhill. I never had an accident.

That love affair is still strong in 2006—the start of my forty-sixth season. Here's how my lobster business evolved.

In 1950 my parents, Roman Joseph and Eleanor Taft (Kew) Prybot, permanently moved from Boston to Gloucester. Situated on Cape Ann, Gloucester is a granite chunk of an island north of Boston that is rimmed by salt marshes, sandy beaches, mud flats, rocky shores and coves and harbors. Complete with a strong fishing heritage, Gloucester prides itself on being America's oldest fishing port. In Gloucester, my parents purchased a more than two hundred-year-old white clapboard house along with a barn and studio for $10,000, all situated on over an acre of land about a half mile from the ocean in north Gloucester. They believed the countryside would be a better environment for my brother John and me (born in 1945 and 1948, respectively) to grow up in. Thanks to the barn and studio, my father could also base his church interior decorating business here.

My Polish-born father, who followed in his own father's footsteps, was one of the few liturgical artists in this country, decorating the interior of primarily Catholic churches. This master held dear to old country ways in America, working under a perfectionist philosophy: either you do something right or not at all. He didn't waste anything. He also loved recreational fresh and salt-water fishing.

My mother, who was of English and Yankee descent, grew up in Keene, New Hampshire. She was educated at Smith College as a landscape designer. She sacrificed her career for a family and was a great stay at home mom. She also had a green thumb.

My brother John, now a librarian at the Sawyer Free Library in Gloucester, inherited my father's artistic talents. I got my deep love and deep knowledge of nature from him.

On summer weekends when he was not working, Dad used to take John and me, then adolescents, fishing "down the rocks" along the rocky shoreline about a half mile from our house to fish for cunners, a common edible inshore bony fish which grows to about twelve inches long. Bare-footed, pacified by an El Producto cigar and protected from the sun by a fedora hat, my father used to fish with a long bamboo fishing pole atop his favorite rock, which had a deep hole in front of it. He baited the single hook with crushed periwinkles.

In the beginning, while he fished, John and I, tanned, barefoot and sporting crew cut hair bleached by the sun, played in tide pools, slipping and sliding on the blue-green algae-covered rocks. Soon my hunter-gatherer instincts emerged, and cunner fishing took hold of me. My dad and I took turns with the pole. I often wondered what roamed beneath the ocean's surface.

More of those instincts surfaced along with a greater love for nature and discovery. From late spring to early fall my brother and I trod in the woodland's vernal ponds and swamps—sometimes knee-deep in muck—and also along the shoreline, especially in its tidal pools. We brought home spotted, painted and snapping turtles, newts, frogs and salamanders, and even giant water beetles as summertime pets. On one occasion, unbeknownst to my mother, I stored a half dozen just-captured green frogs in the top drawer of my bedroom dresser (after emptying out its contents) during the height of frog mating season. Well, once darkness fell, an incessant "chug-a-rum, chug-a-rum" reverberated from the drawer.

Another time, my surprised mother happened upon a two-foot-long common snapping turtle clawing along the edges of our partially filled bathtub. The mailman gave it to John and me after stumbling upon the reptile crossing the road that morning near our house and kicking it into a cardboard box. Our mother immediately made us release the surly animal into a nearby swamp.

More summertime family excursions (most of the time without my dad, who was at work) to Cape Ann beaches and our favorite spot "down the rocks" for picnics provided the opportunity to collect sea creatures including horseshoe crabs and sand dollars. Contrary to horseshoe crab folklore, which says that these primitive arthropods collect and store shiny objects like coins in their chamber-like book lungs, I never once caught one that did have something stored there.

More hunter-gatherer instincts only surfaced. I once constructed a crude two-foot-long seine net out of a piece of sheer curtain, tying in Styrofoam chunks to float the top and pieces of lead to weigh down the bottom, to catch minnows or killifish in one of the tidal pools. The fish were often too fast to catch with your hands. The seine worked. The fish were always released.

By age eleven, I hunted lobsters during the summer along the shoreline at low tide by either listening for the bubbly breathing noises and then overturning rocks to get at them, or spearing the lobsters from the surface from an inner tube. A mask, snorkel and a pair of flippers and a crude spear consisting of a long pole pointed with a sharpened finishing nail made the job easier.

Then, a defining moment in my life occurred in the summer of 1959 during a family picnic "down the rocks." The sight of a nearby skiff lobsterman hand-hauling a wooden lobster trap drew me to the water's edge. Fascinated, I watched him pull the trap over the rail, open the door, measure and peg the lobsters and rebait the trap. Right then and there I said to myself, "That's what I want to do for a living." I soon began pestering my parents to buy me a boat and lobster traps.

I had also grown big and strong enough to bike the mile roundtrip to Lane's Cove, the closest working waterfront, that summer of 1959. Lane's Cove fueled my dream. This granite block harbor, which was built in the 1880s and juts into Ipswich Bay, had approximately fifteen working lobstermen of primarily Yankee or Finn descent. There was George Morey Sr. (who also operated a fish trap in back of the Cove); Duke Torretta (the "one-armed lobsterman"); Herm Marchant; Alec "Tut", Pat and Moe Ahonen; Arthur Gaudreau (the "Frenchman"); Uno Niemi; Edwin Erickson; Eino Leino; and Jack Lutti, to name a few. I quickly got to know them. The lobstermen seasonally worked their strings of fifty to one hundred single traps in either hand-rowed dories or twenty-five- to thirty-foot-long powerboats. A few owned fish shacks down at the Cove. The Cove was a seasonal port, especially since ice cakes often jammed it solid during the winter. By then, most of the boats were hauled out along the Cove's shorelines. The lobstermen stored their traps on the wharfs at their special places.

Although the Cove and lobstering might have looked good to an eleven-year-old, many in the community held social stigmas against the two in the 1950s and 60s, especially since the Cove's west wharf harbored a group of ten to fifteen winos who regularly gathered and drank there during warm weather on a long crude bench under the old apple tree. A lot of townspeople regarded the Cove as a haven for social dropouts. Many of these men nipped down fifty-five-cent apiece pints of white port wine daily to the point of passing out at day's end. Although boisterous at times, the group never seemed to bother anyone then.

The good-natured "Benny the drunk" was one of their best-known characters. After being suddenly awoken out of a deep slumber while passed out in the tall grass by loud voices, hidden in the tall grass, he would suddenly shout out, "Shut up, you bone heads". The last member of the old Cove gang was "Yiksi", who remarkably quit drinking on his own. He died in the year 2000.

My father, along with much of the community, also viewed lobstering as seasonal, uncertain and cutthroat work—in short, a job that held no future. In addition, they believed it to be a last resort for many social dropouts who worked only long and hard enough to buy their next bottle of liquor.

Naturally, I didn't see lobstering the way my father did. With my naïve, youthful thinking, I once hoped he would quit his job and become one of the Cove's lobstermen, not realizing then that he was one of the country's top liturgical artists and that getting there required a lot of talent, dedication and hard work. I guess I also missed my father, since he was away so much. And, I wished that I would get to do more things with him.

"I want to go lobstering," I told the Cove's lobstermen that summer of 1959. They responded, "You damn fool—don't even think of going lobstering—go to school and make something out of your life." First off, the news of a new competitor is never welcome to any lobsterman, then or now. Although I'm sure they viewed me as an outsider and an out-of-towner with no family fishing roots at all and they probably wished I would disappear, the men tolerated me; but most of the men were sincere in not wanting me to repeat their mistakes.

Late that summer, my father, who had the final say in the family, gave in to my nagging. He purchased a twelve-foot, flat-bottom wooden-plank skiff complete with oars for fifteen dollars. His old excuse for holding off, "you have to wait until you are big enough to swim," was no longer applicable. My father also bought twelve new and used wooden lobster traps with buoys and lines. We stored the gear at home for the winter.

Obtaining a lobster license then simply meant sending in a completed application with a five-dollar check to the Division of Marine Fisheries at Ashburton Place in Boston. You picked your own buoy colors. Moorings at Lane's Cove were just as easy to get; the elders told you where to moor.

If you did things their way, everything would be all right. You had to work your way up to the safest berths. The skiffs were tied off the wharfs in the deeper sections where they wouldn't ground out at low tide, while the powerboats were anchored bow and stern in the deeper middle section. Lobstermen ferried to and from their powerboats in tender skiffs. In those days, there were very few pleasure craft.

As soon as school got out for the summer in June of 1960, my lobstering career began, and so did the memories. That day Tut Ahonen transported my freshly painted gray-top and red-bottom skiff, complete with a pair of oars, a wooden baler and an anchor with a line, and my lobster traps, down the Cove with his old Studebaker pickup truck. Ahonen was nicknamed Tut for his short stature resembling that of King Tut.

Trap day for me couldn't have come quickly enough. I gradually rowed out all twelve of the thirty-two-inch-long wooden-lath traps and set these along the rocky shore within a quarter mile of the Cove, having no idea what the grounds were like below; this was all water to me. Each bone-dry trap, baited with redfish, reluctantly sank to the bottom despite three additional soaking bricks. My red, yellow and red buoy colors showed nicely on the mid-morning calm ocean surface. Despite three painful blisters on my hands, caused by the rowing, I could already feel the freedom of the ocean and the job that morning.

The Workforce

Nova Scotian lobsterman Richard Nickerson represents nine or ten generations of fishing in his family.

I waited impatiently for my first haul. After making and eating breakfast around 6:00 a.m., I biked down to the Cove, taking along my boots, oilskins and a pair of cotton gloves. At the Cove, most of the other lobstermen had already headed out to sea.

Hauling up that first trap to the surface was easy, but breaking it over the rail was another story. I made the mistake of pulling the pot into the railing instead of lifting it over. The sheer excitement of not knowing what was in that trap bolstered my strength. After only getting shorts or "crickets" in the first two traps, the third pot finally yielded a two-pound keeper—my very first legal lobster. In no time, handling the very lively shorts and pegging the legal size lobsters' claws with wooden pegs quickly became second nature. I stored the day's seven-lobster catch—about ten pounds—in a special holding crate (or lobster "car") and tied it off my stern mooring. My summer lobstering routine consisted of hauling the traps early in the morning, getting bait for the next day and then selling the catches to a restaurant at week's end. The bait usually disappeared overnight thanks to the cunners. I always took Sundays off and went to church.

Part of my lobstering sea time was spent watching and learning from the other lobstermen, especially Herm Marchant. Marchant, who always dressed in a plaid shirt with ragged sleeves, a fedora hat, a pair of rubber boots and Black Diamond oilskins, worked his fifty half-round traps in a sixteen-foot-long slate gray and red dory. One morning I made the mistake of offering to help the senior lobsterman haul some of his pots. He snapped, "No!" I never made that mistake again. I now realize that Marchant didn't want anyone else to touch his lobster gear, and that I had also insulted his independence.

George Morey and his summer helper, Bud Brailey, a schoolteacher, operated a fish trap in back of the Cove. Such fish traps, which seasonally targeted mackerel, herring, squid and butterfish, were common sights along the Cape Ann shoreline then.

I used to get mud hake and small pollock lobster bait from the fish trap. Mud hake, or squirrel hake, and pollock were two common schooling fish species that regularly swam into the fish trap. Other times, I hand-lined mud hake evenings or trapped cunners during the daytime with special cunner traps or I went to Gloucester with Tut to one of the city's many redfish processing plants. Tut also took my lobsters to market for me—which was then a seasonal seafood restaurant that closed after Labor Day and reopened around Memorial Day. Lobsters were worth about fifty to fifty-five cents a pound to the lobsterman back in 1960.

One day my tolerance to seasickness was tested by a moderate swell from a previous night's northeast wind. Rowing home that morning, I suddenly felt hot, light-headed, weak and clammy. These seasick symptoms all erupted into vomiting, followed by a brief period of instant relief. Setting both feet on solid ground provided the cure that day. Time has healed this ill to the point that I'm not bothered by it—at least not on my own boat.

One day during my first summer of lobstering, I got to be the lobsterman hauling traps in front of my family's picnic spot. That day my brother, mother and a good friend of hers with her three children held a cookout there. Not only did I supply the lobsters, but I also took some of the people out to show them about lobstering. I was so proud.

By Labor Day and the start of school, I had already hauled out and stored at home my waterlogged traps and skiff nicknamed "the wedge" by other lobstermen. The first season left me more prosperous and knowledgeable about lobstering and business. "When are you going to pay me back the forty-five dollars for the boat and traps?" my father soon asked. His request shocked me, as I had believed that the boat and gear were gifts. I ever so reluctantly paid him in full. My father knew what he was doing. I now realize he taught me fiscal responsibility and that if you work hard and pay for something yourself, you will appreciate it to the fullest.

My hard work and determination also earned me the respect of the other Cove lobstermen. They were always helpful and kept a watchful eye over me. One of the old-time Finn lobstermen, Alex—partially crippled by a stroke—even taught me how to knit lobster trap "heads" that summer. Trap heads are the nettings within the trap. Having difficulty pronouncing my last name, Alex called me Trypod. But Alex, like most of the other lobstermen, kept reminding me, "It's okay to lobster now, but don't do this for a living."

The Workforce

My lobstering quickly became more than a summer livelihood; it became a way of life that extended into the off-season when I built new traps and repaired old ones in the dank basement of our house. I wanted to increase my string size to twenty-five traps. The old worm-eaten lath traps were given either the standard thumb or hammerhead pushdown test. A piece of wood that broke simply needed replacement, often by softwood laths. Very rarely was an old trap thrown away. "The old junks fish the best," reasoned many of the lobstermen. For new traps, I nailed together oak trap kits purchased at the local S.R. Harvey and Sons sawmill for $1.80 each, starting with the frames and ending with the laths, bricks and netting. I also made some homemade half-round traps using bent oak saplings cut from the woods for bows and old plaster laths from demolition sites.

During the spring I used to set a few traps off the wharves at Lane's Cove and also at Folly Cove, biking there and hauling the traps before school. The few lobsters caught were stored in a crate at Lane's Cove.

My lobster business only grew—more traps, bigger skiffs with outboards and even Briggs & Stratton-powered pot haulers. The new rigs increased my fishing ranges to more productive grounds. Summers also saw me building and setting more oak traps. I even used my thirty-six-inch English bike to individually transport new traps to the Cove. Placed flat against the frame, each pot was tied to the top of the frame while the underlying down-positioned pedal bore its weight. I simply walked the bike with the trap on one side to the top of the hill and then balanced myself on the other side, sitting sideways on the seat with my hands on the handlebars and brakes, gliding the balanced vehicle all the way downhill to the Cove. There was never a single accident.

Once I became a teenager, I'm sure my mother and father had mixed feelings about my deepening involvement with lobstering. No doubt they were grateful that lobstering offered a constructive physical outlet for my abounding energy; yet they worried that it was robbing me of my youth, particularly as my father saw his hopes of me following in his footsteps further vanish. "Why did he have to get involved with that damned lobstering?" my frustrated father repeatedly asked my mother.

But my dad's hopes were briefly buoyed in the fall of 1961 while he redecorated our own parish church in Lanesville, Sacred Heart Church. Every afternoon of that fall I got off the school bus in front of the church and helped my father and his three-man crew for about ten weeks with washing, spackling, painting and even stenciling, often high up on scaffolding.

"Some day you will remember me by these," my father told me as I assisted him with gluing on his two circular canvas Christ and Blessed Mother murals high up in front of the church. How true his words proved to be. Sacred Heart is still my family's parish church, and every Sunday I now look up at the murals and think of him and our experience. Fortunately, these murals still remain after two church repaintings. Although this work experience was very enjoyable and seemed to come naturally, it did

Here, lobstermen at the annual Maine Fishermen's Forum at Rockport, Maine, stop by the Brooks Trap Mill booth. This company, headquartered at Thomaston, Maine, sells new wire traps as well as fishing supplies.

not break lobstering's grip on me. The shortage of priests and money has since forced the closure and sale of Sacred Heart Church.

Once I turned sixteen, I got my driver's license and a set of wheels: a secondhand half-ton Chevy pickup truck painted fire engine red. Lobster earnings paid the entirety for the $1,200 vehicle. The lobster earnings also paid for my approximately $1,500 a year college education at University of Massachusetts Amherst where I earned a B.S. in Marine Fisheries Biology in 1970. By college, I had built up my business to a half-ton pickup truck, an 18-foot outboard-powered skiff with a pot hauler and a 250-trap string.

I quickly learned that college, like life, is largely what you make of it. Combining the textbook knowledge, especially that of the advanced biology courses—ichthyology, vertebrate and invertebrate zoology—with hands-on experiences made the material all the more meaningful and easy to remember. During college I took up scuba diving and began making winter weekend and vacation day trips out on Gloucester draggers to better explore Cape Ann's waters. I began writing about and photographing the fishermen then, too.

Since fishery biology jobs were hard to come by in 1970, I decided to continue lobstering, now being able to extend my season into October. A college education was not the panacea that my parents, fellow Cove lobstermen, and even myself thought it to be. College was merely another steppingstone in life, which made me realize you have to keep moving and create your own luck.

The Workforce

Wanderlust got the best of me during the off-season of my first extended lobster year. That November and January I made several ten-day-long offshore fishing trips as a guest aboard the 117-foot-long Gloucester dragger *Judith Lee Rose* with her seven-man crew. I took photos and gathered knowledge and writing material on the fishermen at work. The wooden vessel often dragged up huge lobsters, sometimes weighing 25 pounds apiece, in depths often over 1,000 feet in different outer Georges Bank canyons. These trips led to my first article, "The Log of the *Judith Lee Rose*," being published in the *Gloucester Daily Times*. I still vividly remember that wonderful feeling of getting that first article published. Months later the Maine-based fisheries trade journal, *National Fisherman*, published the results of my senior fisheries biology thesis on the Gloucester fishing fleet.

That winter I also traveled to the West Indian island of Barbados and to Peru in South America. I visited Machu Picchu, the lost city of the Incas in the Andes. The urge to explore the world and its people was just as strong as that to explore the ocean and its fishermen. Combining lobstering, writing and photography and traveling seemed like a good life style option then.

But that winter of 1971 the University of Massachusetts opened up a marine station at Hodgkins Cove—only two miles from my house. Then directed by renowned scientist Dr. Charles S. Yentsch, the lab's research largely focused on the Gulf of Maine's phytoplankton link in the food chain, or primary productivity. Dr. Yentsch's wife, Dr. Clarice M. Yentsch, also worked there along with about twenty other staff including research scientists, graduate students, lab technicians, clerical workers and Captain Bill Murphy. A Newfoundlander and former fisherman and the retired captain of the National Oceanic & Atmospheric Administration research vessel *Delaware II*, Murphy skippered the lab's fifty-foot research vessel *Bigelow*. The station also offered a special "summer science enrichment program" for elementary school students interested in the ocean.

I landed a 9-to-5 job there in 1971 as a combined lab technician and a teacher in the summer science enrichment program. This was a tremendous opportunity that I couldn't let pass by. I also sold most of my lobster traps, except for a skeletal string that I fished after work and on weekends with the eighteen-foot skiff.

Despite the lab being a great place to work, also offering many fun social activities and the chance to meet people from all over the world, I resigned from the job in the spring of 1973. It just wasn't my calling in life. No doubt my parents hoped I would meet a nice woman at work, settle down and get married. I was still married to my lobster business then.

Being a free man and running my own business again felt good. Naturally, I had to rebuild my lobster trap string. That winter I also purchased a second-hand forty-foot Novi (Nova Scotian-built) longliner, powered by a 135-horsepower Ford diesel to do some off-season tub trawling in the Gulf of Maine. Longliners snag fish either on or off the bottom with baited hooks affixed to long ground lines. With a sale price of $14,000, this was the first time I had ever taken out a bank loan. I renamed the beamy and beautiful Cape Islander vessel *Machu Picchu*.

On my first day trip in the spring of 1973, approximately 2,500 pounds of cod and cusk pocketed me about $450. Once the dogfish arrived in May, I resumed lobstering with the skiff. That summer I also took my father and terminally ill mother and a group of friends on a very special day excursion aboard the *Machu Picchu* out to the Isles of Shoals, about eighteen miles north of Cape Ann. My mother had already endured two cancer operations followed by radiation treatment within the past seven months and was now in chemotherapy. We toured Star Island and had a picnic there. After a long courageous fight, my mother died in May of 1973. Losing a loved one for the first time was hard for me—especially my mother, the very person who brought me into the world and raised me through all the good and bad. I loved her very much.

The *Machu Picchu* soon became a financial burden. One thing after another broke down, and I still had an old boat. Plus, the boat was too big to fish alone. I soon realized that big was not necessarily better and that I was an inshore one-man-crew skiff fisherman. I sold the *Machu Picchu* at a loss two years later—just glad to get rid of it.

Being young, I quickly bounced back. Soon, a new eighteen-foot Norton fiberglass skiff with a self-bailing cockpit and spray hood then only costing $2,200, along with a fifty-horsepower outboard and a hydraulic pot hauler, replaced the Novi skiff. I named the deep blue-hulled skiff *October Sky* after my favorite month of the year, known for its bright blue skies. This fast and roomy boat now allowed me to fish more traps and set in five-pot trawls. I could also tub trawl on many of my old spots in Ipswich Bay, all without the big boat headaches.

Two big moves in my life occurred during the mid-1970s: one was moving out of the old homestead and into an apartment at Pigeon Cove Apartments in Rockport—about two miles away. I simply needed some distance from my father and probably the same was true for him. I still checked on him every day to make sure he was all right.

The other move was to Pigeon Cove Harbor for a new mooring site. This granite block harbor behind the Cape Ann Tool Company, which had hoists and didn't freeze during the winter, was much easier and safer to work out of than Lane's Cove. About thirty full-time lobstermen worked out of here.

During my off-season starting late fall when the boat and traps were also hauled up, I resumed world traveling, which included a three-week-long wildlife photographic safari in February to Kenya and Tanzania. Now instead of hunting and catching lobsters, I was bouncing about East Africa in Land Rovers tracking down and photographing elephants, lions, giraffes, hyenas, hippos and leopards. When not traveling, I made daylong fishing trips aboard fellow Pigeon Cove Harbor lobsterman and good friend Jerry VanDerpool's forty-foot Novi *Peanutbutter Pat*. "Van" and I used to quick-set five tubs of trawl gear baited with mackerel chunks on Tillies Bank about twenty-five miles off Cape Ann, often catching three thousand-pound trips of big cusk and haddock in the spring. My lobstering typically started up around May 1.

In October of 1980 I met a wonderful woman, Anne Medico, who was living nearby in Rockport. We hit it off well right away and began to seriously date. On January 9,

The Workforce

1982, Anne and I got married in a traditional wedding ceremony at St. Ann's Church in Gloucester, later honeymooning in Italy and East Africa. The wedding day was an especially joyous event for Anne's parents, Ernie and Louise, and my aging father, who lived long enough to see me get married. My brother John was also home and served as best man after being away for nearly a decade serving with the Peace Corps in Guatemala. Sad to say, that following August my father died from liver cancer at home in his bedroom, as he wished. After this Anne and I moved back into my old homestead.

I'm still a skiff fisherman, but since the mid-1980s I've lobstered year-round to keep the cash flow continuing. People ask, "How do you do this during the winter in an open boat?" You simply dress appropriately to keep warm, and this often includes putting on thermal underwear, a facemask, a scarf and taking along an extra pair of thermal-lined rubber gloves. During bitter cold winter days, if either your hands or feet get wet, then you get cold all over. My wife and friends are working on me to cease winter fishing.

In 2004 I bought a new Southshore twenty-foot-long fiberglass skiff complete with a molded top. Webber's Cove Boatyard in Blue Hill, Maine, built the Calvin Beal Jr., designed craft. I love it. Powered by a ninety-horsepower Honda outboard, it's fast, over thirty knots, while also sea-kindly and able to carry a good load, and it can turn on a dime. I feel like a cowboy out there now.

I'm still married to the all-consuming lobster job and way of life, despite having set more limits and realizing there's more to life than lobstering. Christmas and Easter are the only holidays I take off, and I haven't taken a sick day yet, even though I've hauled gear feeling sick many times. I still work under the principle that I'm done when the job is finished. I'm one who pays his bills, too. I have downsized my strings from 700 traps to 375. The days of hauling 200–300 traps a day alone are over for me. The job is continually never done.

Anne and I have already been married twenty-four years; we are the proud parents of a son, Tom, born in 1989. He has already started a landscaping business. Tom does go out lobstering with me on occasion, too. My family still gets to go on an occasional trip, especially to Disney World, and we still do family cookouts "down the rocks."

There have been ups and downs with the lobstering off Cape Ann, but the lobsters have kept coming and I've made a living every year. Today I still wouldn't trade lobstering for any other job in the world, and if I had to do my life all over again, I wouldn't change a thing. In what other occupation can one simultaneously earn a living, be one's own boss, work close to nature and be in sync with her rhythms and seasons, experience solitude (most of the time) and also get fresh air, exercise, and your own seafood?

I generally work amongst a great bunch of guys. Sure, there are a few cutthroats, but most are hard workers who tend to their own businesses. In times of need, lobstermen will come together and give you the shirt off their back. Here's one story that illustrates the nature of some lobstermen. Out lobstering one summer day, the late Roy Gray, a tall and lanky, good-natured Pigeon Cove Harbor lobsterman and a lifelong bachelor, had put his day's catch into a lobster car (or crate made out of wire) and had set this on his railing while cleaning out his boat at day's end. Well, a wake rolled his boat several times, and you guessed it, knocked Roy's crate of lobsters overboard, quickly sending everything to the bottom. He cursed and swore and headed home.

While docking, he told other lobstermen, including Jack Ketchopulos, of his ordeal. Later that same day, Ketchopulos had brought along a scuba diver to retrieve a lost trap. Yes, to everyone's surprise, Roy's crate had landed right alongside that trap, and the diver salvaged both of them. Ketchopulos didn't tell Roy about this right away; instead, he tied the crate to Roy's railing—the same way the bachelor lobsterman always did. Well, Roy came down that afternoon, saw the crate hanging from his boat, and he couldn't believe his eyes. Most still remember his comments: "I must be losing my mind; I thought I lost that crate overboard today." The kind-hearted prankster Ketchopulos soon told him the real story.

My five-minute commute to work (about an hour before sunrise), rarely seeing another vehicle, is a job amenity that would make most commuters envious. I still make my own luck and go after opportunities, not waiting for them to come to me. The world is full of opportunities. You have to be willing to work. The opportunities just need to be either created or discovered. I have learned you buy supplies when you don't need them and that most breakdowns occur on holidays, and that despite improved weather forecasting, the weather is never etched in stone. I find myself fishing smarter and not harder, no longer wanting to catch every lobster out there, but just a fair share. I always prepare to go out lobstering the next morning. I still like to travel to other ports and "talk fish" with their lobstermen.

Plus, (if you make the time) one can enjoy some of the industry's seasonal added treats such as the Maine lobster boat races in the summer, the Maine Fishermen's Forum and Trade Show, the Massachusetts Lobstermen's Association annual weekend and lobster trade show in the winter, the International Boston Seafood Show in the early spring and Fish Expo and Workboat Atlantic, which lately has been coming to Rhode Island every other year. This is still the best job in the world.

II
An Inshore Lobsterman's Year in Review

The next two chapters, buoyed by personal journals, explore what a lobsterman might experience at sea on a season-by-season basis. Out there, while behind the wheel and at the rail of our boats, we spend a lot of time looking both down into the water and around, especially toward the horizon. Besides astronomical constants like tide and moon cycles and the seasons, lobstermen experience all kinds of weather and sea states, regular and surprise biological events, and even changes in their moods.

Every lobsterman works a garden patch in the Atlantic. For some, this plot is miles long, hundreds of feet down, and out of sight of land. My garden in the southernmost Gulf of Maine runs along an approximate five-mile stretch of northern Cape Ann's primarily rocky shoreline out to the man-made Sandy Bay breakwater, about two miles off of Rockport. Much of this territory is very vulnerable to easterly storms. Seaward, that rocky stratum gradually phases into sandy and muddy bottoms about sixty- to eighty-feet down. That bottom gradually descends to three hundred-foot depths and is broken up by ridges and humps. Most of my year-round lobstering is done in the shallows—"in the rocks" or in "the bushes"—and is rarely done deeper than eighty feet down.

Spring

Spring, the season of rebirth, is a good starting point for the year in review. The natural annual flow in our temperate zone, which keeps the marine ecosystem going, is for an awakening in the spring, big production in the summer, followed by a wind down in the fall, and lastly rest and little activity in the winter.

The start of another astronomical spring occurs in the northern hemisphere around March 21 (the official vernal equinox) when the sun shines directly over the equator, sets due west and rises due east, gives equal amounts of daylight and darkness and gets higher in the sky. The March sun is powerful and warm. The mornings are suddenly light. The change to daylight savings time occurs around the last Saturday of March, and we "spring ahead" one hour. The daylight and air, water and land temperatures gradually increase, too, triggering the terrestrial and marine productivity machines. The nutrient-rich and rested ocean water of winter quickly comes to life with intense feeding, moving, breeding and growing activities.

Spring gives the barometer a workout. Weather can make or break a lobsterman's season. The season's weather often gets stuck in patterns, ranging from extreme warmth, cold, dampness with lots of fog, and storminess with lashing rains and northeast gales, to mixtures of all that depending upon the jet stream patterns. These northern and southern atmospheric oceans of air allow a continuous bombardment of high and low pressure weather systems along with their warm and cold fronts to move across the country, up the coast, or down from Canada, and determine local weather. Regardless of how cold the spring might have been, the lilacs always seem to bloom on or before Memorial Day, and everything averages out by summer.

The looped jet stream pattern frequently brings up moisture-laden storm systems from the south and often positions these just off the coast to form the northeaster or "line" storm, as the old-timers called them. The northeaster, the most feared of storms, which has been seen in forms such as the Portland Gale, the Blizzard of '78 and the No Name storm, form off the coast and usually move up the coast or in a northeast direction and give northeast winds. History has shown that northeasters can take many different courses: they can stall and intensify while moving up the coast, or fizzle, or even sometimes retrograde back to the shore instead of either moving out to sea or further up the coast. These storms frequently strike during the equinoxes and solstices and the monthly astronomical high and low tides.

The lengthening days and rising temperatures plague lobstermen with spring fever year after year. When spring has sprung, you can smell, hear and feel it in the air. Smiling and talking become second nature again, and hope and enthusiasm, especially for a banner fishing year and good weather, flow anew. Just some warm weather and a sign of a few lobsters will trigger many lobstermen to set their traps, and "The gold rush is on again," says Fred Hillier, a veteran Pigeon Cove Harbor lobsterman. These lobstermen fear their competitors might get a jump on them if they don't set their traps.

This season can be a very anxious time for the lobsterman. First of all, more money frequently goes out than comes in then. Under that pressure, the harvester must decide, do I set my traps early and risk possible storm damage, or do I wait and possibly miss the spring run? It is always a tough decision. So much of the spring fishing is done in the shallows, which are prone to damaging traps. Spring, with its fickle weather, is often the hardest season on the lobster gear.

For those lobstermen like myself who leave gear out all winter, live-caught lobster production is usually the lowest in the early spring due to cold-water temperatures. We find ourselves changing a lot of water in the traps this time of year.

In addition, spring always brings a heavy workload, including painting the boat and the buoys, repairing and updating the gear (replacing trap tags and biodegradable hog rings, and now having to make the lobster gear whale-safe), and setting it. Lobster buoys must have a breakaway link, and beginning in 2007 for some of the fishing areas, floating ground lines between traps can no longer be used.

Numerous plant and animal harbingers help lobstermen notice signs of spring at sea and often along the immediate shoreline. By early April the first diatom bloom occurs, and millions of microscopic green glittering primary producers briefly turn the navy-blue waters of winter green, only to bust by a combination of being grazed upon by zooplankton and depleting the nutrients in the water themselves.

Then knotted wrack or rockweed, *Ascophyllum nodosum,* "goes spawny" along the rocky shoreline's intertidal zone. During this roughly month-long process, which usually begins in mid-to late March, the algae's olive-colored fronds turn a slimy brown, and their tips' yellow receptacles first swell up and then break away, soon releasing gametes for the future generations.

The flowering of certain shoreline plants such as shadbush, forsythia and the seaside rose announce spring, too. The forsythia bush, with its cadmium yellow flowers, leads the way in mid-April, followed by the shadbush's mid-May white flowers, and lastly by the fragrant white, magenta and pink seaside rose flowers of early June.

Incidentally, many cod fishermen associate forsythia's blooming with the presence of codfish in Ipswich Bay, a major spring cod-spawning area. "When the forsythia is in bloom, there's codfish in the Bay," say the old timers. The final floral messenger is the pine trees casting free their pollen in early June, which also causes me to uncontrollably sneeze.

Animal harbingers catch my eye, too. Late March is announced by the white specks of once-planktonic barnacle larvae suddenly settling on new fixed and floating homes—especially buoys, traps and rocks. The approximately two-week-long intertidal sea worm spawning quickly follows. Stimulated by the flowing tide, these sexually mature annelid worms, greenish-gray in color, leave their burrows in the sediment to deposit eggs and sperm in the water, often turning it milky. Gulls know the routine. Most of the worms barely have time to spawn before the gulls gulp them down.

The two- to three-week-long inshore spring lobster run usually heralds late May or early June. Many of these lobsters (mainly large females with rock-hard, dull-colored shells, some even fouled with barnacles) migrate toward the warmer and safer shallow water to first hatch their ripe eggs and then molt and mate. Other females move inshore to extrude their internal eggs to the undersides of their tails and become the "black eggers," which are common inshore in the late summer and fall. Some years the spring run has bypassed the inshore waters of Cape Ann.

Suddenly catching good numbers of two- to five-inch-long rock eels and cunners in the traps again also tell me it's spring. These fish then move inshore and become active again. The rock eels often drop out of traps that were recently hauled aboard onto the deck, where they wiggle with serpentine motion. A bottom water temperature above fifty degrees Fahrenheit, often felt in May, seems to trigger the cunner movements. Usually the spring lobster run is not too far away from the first sizeable sighting of cunners in traps.

Bird movements herald spring, too. By May, the wintering loons, mergansers, murres and the colorful and acrobatic harlequin ducks have headed north to breed. The loons fly inland to freshwater bodies to nest.

Songbirds also briefly flutter above the ocean in the spring during their east coast flyway northern migration. Exhausted, thirsty and often lost in the fog, warblers and vireos have landed on my boat to rest. Seagulls pluck these tiny birds out of the air and swallow them in one gulp.

The sight of a resting flock of migrating blue herons at about the same time and place along the shore every spring tells me it's late May. These three- to four-foot-high bluish-gray, long-necked, long-legged and long-beaked pterodactyl-like birds are very skittish.

The final spring harbinger, the April departure of most of those cute but pesky and pilfering harbor, grey and hooded seals to more northern waters, is welcomed by Cape Ann lobstermen. Some three hundred seals winter on the Big and Little Salvages, the rocky outcroppings and ledges about three miles off of Rockport. The dried fur of seals basking in the sun often turns the dark Salvages's shoreline white with seals. The big eyes and whiskers on the seals' faces constantly remind me of my black Labrador retriever's face.

A number of biological surprises have occurred throughout the springs. In June 1999, the pine trees' airborne sulfur-colored pollen was so thick in places on the ocean's surface that it resembled a freshly spilled fifty-five-gallon drum of yellow paint.

The next year during early spring, a huge population of edible mussels was born. These one-eighth-inch-long bivalves apparently swarmed like bees and completely fouled the insides and outsides of many lobster traps. Some lobstermen either pressure-washed or brushed free the black specks at sea only to find their traps completely re-fouled just two to three days later. Fortunately, the swarming stopped after a month.

Who can forget the sea urchin population explosion of the early 1980s? These echinoderms devoured much of the kelp and the Irish moss growing along the rocky coast's sub-tidal zone, later turning these areas white. The urchins even went after the plants growing on the traps and the bait inside, sometimes chewing out the traps' netting. The urchins filled the traps so full that I had trouble breaking the four-foot-long wire traps over the rail. Many lobstermen, including myself, suffered painful, long-lasting hand and knee sea urchin spine infections. Luckily, their population crashed after the urchins exhausted much of their food supply.

One of the biggest coastal lobster runs of mainly two-to-five-pound females occurred in June of 1991. One thirty-six-inch-long wire trap of mine caught twenty-five pounds of two- to six-pound-apiece lobsters on a two-night set, my highest single trap weight catch. (By the way, no maximum-size lobster limit existed then.) This run lasted about three weeks.

Spring 2005 outdid itself with biological surprises. Not only has there been a kelp comeback, a gargantuan red tide bloom which closed the clam flats from mid-coast Maine to Cape Cod from spring to early summer, and an exceptionally heavy pine pollen dusting, but also a barnacle barrage some area lobstermen say they have never seen before.

The razor-sharp barnacles not only make the pots heavier and more dangerous to handle, but also cut down on their fishing efficiency since lobsters, especially the weaker ones with new shells, do not like to crawl into barnacle-covered traps. Many lobstermen attribute this year's barnacle onslaught to a paucity of the planktivorous herring in coastal waters. "Dense swarms of barnacle larvae in the early spring constitute the herring's principal prey," write Bigelow and Schroeder in their legendary *Fishes of the Gulf of Maine*.

The lobstermen have been fighting back by either docking and then cleaning and resetting their contaminated traps, or by leaving the fouled ones ashore and replacing these with clean ones, or by scraping and or pressure-washing off the barnacles at sea.

Summer

Much happens between this season's June arrival and September departure. The marine ecosystem's productivity machine makes hay while the sun shines under ideal conditions of abundant light, food and warmth. Flora and fauna grow, build up their body reserves, and multiply as new generations hatch and old ones die.

The astronomical summer solstice arrives around June 21, and by then the daylight has peaked with over fifteen hours a day. June 21 has the latest sunset at 8:25 p.m. and the earliest sunrise at 5:07 a.m. The days then seem endless. During this time the sun is furthest north of the equator and the northern hemisphere also tilts toward the sun. The sun is at its highest in the sky, and it sets and rises farther to the north than at any other time of the year. Sunlight now shines through different windows than it did a few months ago.

The days already noticeably shorten after the Fourth of July. By early September, approximately three hours of daylight have been lost. The sun sinks lower in the sky during the day, and the early morning and late afternoon shadows get longer.

The air temperatures typically peak by the third week of July when heat waves also usually occur. Many fishermen refer to August as "the hot month" because the water temperatures are generally at their warmest, frequently in the high sixties to even the low seventies, sometimes continuing right into October. Lobstermen often experience "September slumps" with their catches when calm weather prevails and does not stir up lobsters and get them trapping.

Summer weather is also never etched in stone. Although it generally quiets down after the summer solstice, the weather still often runs in extremes and gets stuck in patterns. It frequently becomes erratic again by September 10, the traditional start of the hurricane season. Just about every Labor Day weekend, the surf's up, most of the time thanks to a passing-by offshore storm's ground swells, and most hurricane seasons give at least a scare or two.

Southwest is summer's prevailing wind direction off Cape Ann, often coming up hard in the afternoon. The season's common haze-filled skies frequently create blushing and salmon sunrises and sunsets and the rising and setting suns often burnish golden pathways onto the ocean's placid surface. Thunderstorms, occasionally violent, with brief, strong, gusty, damaging winds, torrential downpours, even hailstones, and intense thunder and lightning, make up the bulk of the summertime storms. Sometimes their cells create "walking rain" over the water as they move along from the northwest to the southeast and drop their precipitation. Many of these storms end with vivid rainbows. Northeasters have occurred in the summertime; I've even had traps wash ashore in July and August from them.

Although the weather is not so much of a worry, summertime creates a new set of anxieties for the lobsterman. There's pressure to make the most money then. The old-timers preached, "If you haven't made it by Labor Day, you never will," although good falls and early winters still bail out a lot of lobstermen. I also feel compelled to start the workday as early as possible to avoid the day's heat and the pleasure boat traffic, and to get to my traps in some of the popular scuba diving areas before dishonest divers do, since scuba diving for lobsters is allowed in Massachusetts.

Other lobstermen, along with recreational fishermen and boaters' activities on the crowded inshore grounds, cause tension, too. You often then get set over by other lobstermen's gear, making it hard to pull up your traps. Fishing inside this time of year can get to be downright "seagull," as everyone vies for the tight inshore grounds, much like seagulls diving for food on the beach.

In addition, a few fellow lobstermen have meltdowns, especially in the hot weather, and hack your gear, usually with sharp knives. The increased boat traffic chops off buoys and drags away traps, and anglers who occasionally anchor into your lobster gear sometimes cut it away in a fit of rage. This boat traffic also sometimes gets in your way when you're setting back your gear.

Many harbingers faithfully signal the busy summer season. The first is lobster-related. The traditional Fourth of July run of brightly colored new-shell lobsters with very sharp spines announces early summer. This run some years has occurred either three weeks early or late or not at all. Next, there's the late-summer presence and dominance of hard-shell and protective egg-bearing females—"the green or black eggers".

The resumption of game fish activities, including striped bass fishermen casting off of the shoreline at dawn and tuna fishing boats racing to the grounds signal early summer, too. Lately the feed, especially herring, hasn't shown up in Ipswich Bay, a once-popular tuna fishing ground, and there has been very little tuna activity there.

By late summer, the inshore waters especially become "very active" with bass and bluefish. These predators bunch up and feed on baitfish, namely menhaden fry, fattening themselves for the imminent annual southward migration. Huge schools of frenzied bluefish, often accompanied overhead by shrieking gulls and terns, boil the ocean's surface red with blood and send pieces of their prey into the air. In addition, striped bass frequently corral baitfish into a tight ring—highly visible at the surface—charge up from beneath and devour them. Occasionally, these predators, along with mackerel and pollock, push their prey into coves, sometimes driving them out of the water onto dry land where gulls will eat them.

Warm air and water-induced bacterial and viral skin infections, including arm and leg boils, rashes and fingernail infections announce mid-to-late summer. Just one break in the skin by either a lobster spine or a redfish bone or a barnacle can start an infection, even one by deadly flesh-eating bacteria. Whenever punctured, I immediately stop working and bleed the wound, sometimes dousing it with bleach or hydrogen peroxide.

Lastly, the inshore arrival of large and harmless Gulf Stream denizens like twenty-foot-long basking sharks and forty-five-foot-long whale sharks, huge leatherback and loggerhead sea turtles, and ocean sunfish, signal mid-to-late summer. I once saw a five-foot-long sunfish, which was probably being attacked by bluefish, propel itself six feet high out of the water twice. This shocked me, since I previously believed these bullet-shaped fish, which occasionally scare people with their shark-like dorsal fin flipping back and forth on the surface, to be very lethargic swimmers.

Summer also has had its surprise biological occurrences, especially plankton blooms. Heavy rains and the dredgings from an Annisquam River project triggered a huge dinoflagellate bloom of *Goniaulx tamerinses*, which causes red tide, in the late summer of 1972. The University of Massachusetts Marine Station brought this problem to light. Filter-feeding clams, mussels and scallops concentrated this deadly top-shaped single-celled phytoplankton's waste toxins to levels high enough to cause paralysis and death in humans, especially if the shellfish was eaten raw. Since this outbreak, clam-flats up and down the coast have been regularly monitored for red tide, especially after heavy rains.

Green fleece, *Codium* sp., an invasive green algae that was probably brought over in a ship's ballast from the western Pacific, arrived in Cape Ann's coastal waters in 1995. Storms now wash these branch-like algae onto the beaches with kelp and Irish moss. Storm action also fills many lobster traps with green fleece.

A first of its kind (to the best of my memory), salp bloom occurred inshore during the late summers of 1998 and 1999. An offshore storm probably blew some of these regular Gulf Stream residents inshore where high water temperatures and levels of critical nutrients triggered blooms. On-shore winds frequently concentrated the gelatinous pellet-size egg chains or "aggs" of these pelagic chordates into fiberglass resin-like consistencies in the corners of coves and harbors. The eggs even clogged boats' cooling system intake strainers. In time, the blooms busted, and cold water quickly killed others.

Another late summer biological occurrence, the shoreline, coves and harbors teeming with huge schools of migrating two-to-three-inch-long menhaden or pogies ("peanut bunkers" as they call them further south), began in August of 1999. This has reoccurred for several years after that. These silvery fry hug the shallows, sometimes swimming amongst the rockweed at high tide, in a vain attempt to escape predators like voracious pollock, mackerel, bass and bluefish.

The biological occurrence that happened mid-summer around 1980 is still vivid in my mind. An extremely toxic microbe, which easily entered your body via any kind of a skin break and would quickly cause severe blood poisoning, flourished at that time along the inshore waters. One week, I had gotten many barnacle cuts on my arms reaching into lobster traps. Shortly afterwards my armpits were sore, feeling like a pulled muscle. That evening while showering, I discovered red streaks running up both arms. I didn't waste any time. At the local hospital's emergency room, the doctor asked, "You're a fisherman aren't you?" before revealing, "You're about the fifteenth one to come in today with that same infection." His prescribed antibiotic quickly squelched the infection.

Fall

Around September 22, the astronomical autumnal equinox happens as the sun shines directly over the equator. Like the spring equinox, this fall day receives equal amounts of daylight and darkness, and the sun sets due west and rises due east, both moving further toward the south by winter. The sun also shines lower in the daytime sky, to the point where it always seems to be in your face, and the morning and afternoon shadows get longer.

Resembling spring, fall is another very active transition season. Air and water temperatures and hours of daylight slowly decline as wind velocities and storminess generally increase, barometers tend to read high, and the wind directions often touch all of the compass points in any given day. When fall's high-pressure systems are overhead with their cool, crisp and dry air, you feel alive.

In the fall Mother Nature signals her flora and fauna to finish up their summer's business and prepare for the upcoming harsh winter, often through migrations to warmer and safer climes. Early fall's high activity slows down dramatically by its official December 21 end. The surface water temperature often drops to the mid-fifties by the end of October, and the forties by December. The thermocline in the water column gets closer to the bottom as the cold works its way down. Eastern standard time begins again during the last Saturday of October when the clocks must "fall back" one hour.

Some fall mornings are so dark that the sky and the ocean are indistinguishable. The cold illuminations of September's harvest and October's hunter moons help light up dark mornings. That fall chill eventually permeates the air once the sun goes down. You seem to really feel the season's first cold. This takes about two to three days of acclimation. Old man winter usually settles in by late November, first with a hard frost and then with a skim of ice on the freshwater ponds and birdbaths.

Autumn's skies, especially at sunrise and sunset, become very colorful and foreboding. "Red skies at night are a sailor's delight; red skies in the morning, sailors take warning" and "Mackerel scales and mare's tails make a lofty ship carry low sails" are omens I take very seriously. Daytime wind scuds (blackish-gray or white wisps of clouds in the northwest sky moving northwest to southeast) which signal the leading edge of a cold front and, more important, imminent wind, also happen frequently during the fall. Inky-colored grotesquely shaped blobs and blotches often partially cloud the eastern horizon during dawn. These clouds, which are frequently highlighted and brushed with pink and orange pastels, inspire the imagination.

Fall weather often runs the gamut: from the warm and dry warm Indian summer stuff, to hurricane threats and actual strikes, to wet and wild northeasters and quick-moving southeasters, to traditional bright, sunny warm days with cerulean skies followed by cold nights with star-clustered skies and blustery days with northwest winds exercising under skies streaked by leaden stratocumulus clouds. Summer's common capillary waves and glassy calms now give way to a hard northwesterly's feather-white close chops or the mountainous waves and swells of northeasters and southeasters, or to the lumpy ocean surfaces created by the tide going into the wind and waves.

The fall shoreline often thunders with pummeling surf and waves. Who can forget the devastating October 31, 1991, December 12, 1992, and December 6–7, 2003, mega-northeasters, the severe arctic cold front of early October 1975, and the 1989 Thanksgiving Day snowstorm which dropped over eight inches of snow?

Fall is the season that simultaneously creates conflicting feelings of relief and anxiety in lobstermen. On one hand Mother Nature gives cool invigorating weather to do things, while also strongly activating hunter-gatherer instincts, subtly saying, "Put aside now or never for the winter, holidays, and income taxes." Yet, on the other hand, she depresses energy levels with her short days.

More fall conflict arises between the relaxing effects of finally having some weather-related days off and the anxiety of knowing you'll probably have twice the workload when you do get out just to catch up. On good days, many lobstermen work "dark to dark" to get caught up. For other questionable days, one must decide, "Do I go, or don't I go?" I make this decision at home by first listening to the weather reports and then looking at my weather instruments. If there's further doubt, I'll step outdoors, glance at the tree branch movements and listen to the sound of the ocean before going to the wharf.

Most of the time I'll go, preferring to get my work done early. If the ocean is too rough, you can always turn around and come home. Many other lobstermen make their last-minute decisions from their warm pickup trucks at their favorite ocean observation spots. The truck heater feels awful good on questionable days. Getting started lobstering is the most difficult part. One lobsterman going out fishing will often inspire others to follow. Before you know it, your day is done.

Autumn also gives many lobstermen an added feeling of relief from early summer's crowded fishing conditions and storm threats in the shallows. The lobster gear is finally spread out as more lobstermen pursue migrating lobsters in the deeper water. Many of those lobstermen then have to worry about getting their gear towed away by mobile fishing vessels who compete for fish on the same grounds that lobstermen trap lobsters.

Autumn's harbingers are plentiful along the shoreline. By early October, the foliage fires of sumac, woodbine, maple and tupelo blaze just up from the shore. Sunshine blizzards of their crimson and gold leaves often rage during blustery days. Simultaneously, early fall's combination of short days and long nights stimulates the flowering of pink, white and purple asters, false ox eyes and goldenrod.

The dinoflagellates (namely *Noctulucia* sp.) bloom in the fall and fire up the dark ocean. The movements of swimming fish and passing boats agitate these top-shaped single-cell microscopic organisms to luminesce.

The growing of *Sertularia* sp., a pinkish-red jellyfish-like coelenterate with tentacles and stinging cells, on the upper buoy lines in September is another sure sign of early fall. You have to be careful not to get the stinging cells in your eyes. The hydraulic pot hauler yanking in the buoy line over the snatch block will often send these tiny animal parts flying in the air. Getting a stinging cell in the eye will cause extreme burning and later itching in addition to creating a bloodshot eye.

The catching of warm-water fishes—triggerfish, filefish, sea horses and scup—in lobster traps signals early fall. Offshore storms and hurricanes usually drive these Cape Ann rarities into northern waters either from the Gulf Stream or up from the south.

A triggerfish nearly snipped off the tip of my finger once. I had put this foot-long specimen in a bucket of water after catching it in a trap. Like its cousin the parrotfish, the triggerfish also has very sharp and large incisor teeth, used for cutting and crushing hard objects. The fish, playing dead, suddenly snapped to life and bit my finger with its parrot-like mouth as I reached in to grab it. I let the angry fish go, respecting its right to defend itself.

The lobster bait quickly disappearing in the traps, often after just two nights, is another sure sign of early fall. A combination of warm bottom water, current action and nibbling cunners are most likely responsible. The cunners voraciously feed during this time before moving off and becoming inactive in the winter.

Early fall's inshore waters become "very active" with bluefish and striped bass as they feed, fatten and group up in preparation for their imminent southward migration. Most of the time, these fish leave Cape Ann waters in early October.

But the pesky seals return to Cape Ann waters from the north in mid-October. The seals migrate in groups and swim like porpoises on the surface, popping their heads up and down as they go along. Once here, they are especially hungry and troublesome, frequently robbing the oily fish baits out of traps, including those set three hundred feet down and ten to twenty miles offshore. One October, seals popped open the doors and stole the fresh mackerel baits out of twenty-five consecutive traps of mine set at the Sandy Bay breakwater.

The return of the loons and the harlequins (in mid-October) and the common murres (the first week of November) to winter off Cape Ann is another sure sign of fall.

The skies becoming peppered with vacillating flocks of migratory eiders, old squaws and scoters ("white wingers," "shell drakes," "skunk heads" and "brownies" as the old-timers nicknamed the different species of scoters) is a traditional September through

An Inshore Lobsterman's Year in Review

Lobstermen like Bob Morris Jr. and his stern man, Joe Roderick, often work in the winter even when snow falls. Here, the men unload traps from Morris's *Spirit* at Pigeon Cove Harbor.

November harbinger. These birds fly head on in great numbers, often just above the waves when the wind blows northeast. Majestic gannets move south about the same time, then are frequently seen dropping out of the sky, plunging into the ocean's surface like arrows after fall-fattened mackerel.

Winter

Although the earliest sunset of the year happens around December 8 at 4:11 p.m., the astronomical winter solstice (around December 20) gives the shortest day of the year and the latest sunrise with just over nine hours of daylight. The sun is then at its farthest point south of the equator and the earth's northern hemisphere tilts the farthest away from the sun. This positioning makes for a very low-in-the-sky sun, which creates extremely long afternoon and morning shadows, and rich colors. The winter sky at dusk during clear days is often tinged with pink. This time of year is also responsible for the sun setting nearly southwest and rising nearly southeast. Already after the winter solstice, the daylight pendulum slowly but surely (often by a minute or two per day) begins to swing toward longer and brighter days, first in the afternoon and, later on, in the morning. By March 1 the days have already gained over two hours of daylight.

Winter ends the annual seasonal cycle. The water, like the creatures that live within it, usually rests; just surviving is now the name of the game. Except when agitated by waves, the ocean water becomes very clear, sometimes down to a depth of over fifty feet. It also becomes uniformly cold or isothermal from the surface to the bottom, often hitting thirty-four to thirty-six degrees Fahrenheit. The lowest light levels and (usually) the coldest air and water temperatures slow down nature's production machine to a near halt except for the spawning activities of cod, winter flounder, herring and northern shrimp.

Diver friends of mine, Andy Arnold, Tim Donovan and Fred Schrafft, all from Gloucester, recorded a thirty-four degree Fahrenheit bottom temperature fifty feet down off Cape Ann during February of 2005. Seawater freezes below thirty degrees Fahrenheit. That February, I picked up just-caught lobsters with my bare hands. The cold came right out of these lobsters. Picking them up felt like picking up blocks of ice.

Winter weather often makes up for the low biological activity. There have been mild winters, cold ones, rainy ones, snowy ones and mixtures of all of them. I judge winters by the amount of ice I take on the railing hauling lobster traps.

When I was a kid, winters seemed to begin in November, then had an early January freeze followed by a late January thaw, and lastly seemed to quit around spring. The seasons were decisive. The coves and harbors often got clogged with ice floes during the January freeze, while the streets and streams frequently flooded during the thaw.

Decisive seasons are the way nature intended it, but things don't always work that way. I can remember recent mild winters on Cape Ann when roses have bloomed in December and January, and the flora and fauna don't know if they should be dormant and hibernating or active. Decisive winters are good for the human psyche, too. By spring, people are ready for change. I've also found good spring runs of lobsters often occur after harsh winters.

Who can forget the one hundred-hour-long February storm (24–28) of 1969 when Gloucester called in the army to plow out the storm's more than twenty-five inches of heavy wet snow? The largely unpredicted Blizzard of '78 was a classic case of extreme winter weather that brewed up a hurricane-force northeaster during astronomically high tides. This storm took the lives of the Gloucester pilot boat *Can Do*, devastated the shoreline and dropped two feet of snow. Repetitive December and January snowstorms created four- to five-foot-high snow banks in 1997, but the winters of 2001 and 2002 were dry and mild; red roses in the backyard even blossomed in February. January of 2004 went down in the history books as one of the coldest Januaries ever, which even saw a negative nine degrees Fahrenheit reading one morning.

Two winter weather foreboders often catch my attention. The first—a halo around the moon or sun known as "sun dog"—forebodes stormy weather, especially if accompanied by no wind and a high barometer reading around 30.2. The moisture of an advancing weather system creates the halo as the sun or moon shines through it, and the air feels very damp and penetrating. The Blizzard of '78 had a pronounced sun dog the day before, and the air had a very cold chill to it.

The second foreboder, which is seen at dawn as a smoky cloud band on the eastern horizon, signals an imminent smoky sou'wester. This usually happens when one first

An Inshore Lobsterman's Year in Review

Big northeast storms often strike during the winter, like this December storm in 2003 which nearly swept Tony Santo's lobster boat *Kiddo* out to sea from Pigeon Cove Harbor. The boat was saved.

wakes up to the three C's—clear, calm and cold—of a high-pressure system that is briefly directly overhead. As the system moves off, its backside winds will freshen out of the southwest by late morning or early afternoon.

By early winter, the lobsterman's mind is working on relaxation despite being enveloped by this season's early-on gloom and doom ambiance created by the short days and dark mornings. I still feel pressured to earn as much as I can then, knowing the holidays and tax time are just around the corner, and every little bit helps. Fishermen must have their taxes filed by March 1.

During past winters, I've only left a skeletal string of about 150 traps to keep the cash flow going and to give my boat and body a weekly workout. The pressure of going out and getting bait regularly stops, and the weather is not so much of a concern since those traps are set in safe areas and the boat has a strong mooring. One of the few joys of winter fishing is cashing in on the steadily rising lobster prices, which generally apex in late March or April. A bucket of lobsters (around 25 pounds) can be worth over $150 then.

People often say to me, "Fishing winters must be awfully cold, especially from an open skiff." I've hauled gear in sub-zero temperatures when the lobsters would quickly

A winter northeaster scene at Pigeon Cove Harbor.

freeze if I didn't get them right out of the traps and into water. You do what you have to do and dress accordingly to keep warm. Doing so often involves wearing thermal underwear, a facemask, scarf, thermal-lined gloves and boots, and dressing in layers. Wet gloves can quickly lead to frostbite. I've had mild cases on several fingertips. I feel the cold more as I get older. My new boat with a molded top offers more protection from the elements.

Once past the holidays and tax time, winter finally gives me some time to catch my breath and do something other than lobstering: vacation, ski, skate, read, attend trade shows and, most important, become a couch potato near the woodstove and nap. The woodstove's radiant heat gets right down to the bone.

Two winter harbingers related to the cold and dry air of high pressure systems tell me the season is here. First your lips chap, nails break easily and skin gets dry, itchy and sometimes even splits into painful cuts. Also, the dry, clear and cold air creates excellent visibility as well as distortion of images on the horizon, especially distant fishing vessels.

The winter of 2001 held two surprise seabird biological events. In late December of that year, offshore storms blew ashore scatterings of dovekeys off of the Rockport shoreline. Many of these birds normally winter offshore. Once washed ashore, these eight-inch-long black and white oceanic birds do not know how to get back in the water, and they can only become airborne taking off from the water. I've picked up such stranded dovekeys before on beaches and the rocky coast and have even put one in my truck's glove compartment to transport it back to the water in a sheltered area.

In January of that year fellow Pigeon Cove Harbor lobsterman Tony Santo saw a pair of colorful common puffins perched on top of the Dry Salvages off of Rockport. Those two were probably stragglers from the closest puffin community, Machias Seal Island. "Their red-colored bills highlighted by the setting sun were spectacular," says Santo.

III
Rare Personal Experiences

Every lobsterman's career includes rare personal experiences ranging from sinkings, fires, stormy weather (especially squalls), near-drownings, rescues and unusual sightings and catches. The last chapter explained what this inshore lobsterman often experiences throughout the year. This chapter looks at fourteen one-of-a-kind memories that have remained frozen in my mind.

The first three involve sea creatures caught by lobster traps in most unusual ways. One summer morning of 1969 I hauled up a wooden lath trap from the muddy bottom forty feet down about a quarter mile off Lane's Cove. I noticed that the trap had a strange tail protruding from its backside. "Could this be a mermaid?" I joked to myself. No, the tail belonged to a four-foot-long monkfish or anglerfish. This flattened, tadpole-shaped bottom-dweller got its teeth stuck in the rear corner of the trap while attempting to swallow it to get at the bait and catch.

These scale-less fish, which have huge mouths armed with rows of inward-pointing conical teeth, typically lure prey close to their mouths by dangling their modified first dorsal fin—the illicium—before literally inhaling their quarry whole. Monkfish grow to be about four feet long. This chocolate brown-colored fish was released unharmed after I carefully pried its teeth free of the trap.

A one-clawed fourteen-pound male lobster was a second creature strangely caught by a trap of mine one August day of 1982 in the sixty-foot rocky depths off Halibut Point. Somehow this creature got its huge crusher claw entangled in the wire trap's gangion, the short line that attaches the trap to the ground line. The lobster came up on the outside of the front part of the trap rather than inside of it. Remarkably, the rope twisted around the lobster's knuckle didn't break the claw off. Once the trap broke the surface, I reached down, braced the lobster and soon successfully brought it and the trap aboard. This male had a very hard black-colored shell with several barnacles on it.

This third story, which occurred on August 6, 1980, tops the bizarre ways my lobster gear has inadvertently nabbed creatures. That morning, I was hauling a string of wire traps

Rare Personal Experiences

set in four-pot trawls about a quarter-mile off Halibut Point, when I came upon a buoy line (with a buoy still attached), badly twisted, kinked and snarled with another lobsterman's buoy and line. Incidentally, the trawls are buoyed at each end. Perplexed, I pondered, "What could have done this?" Getting caught in a boat's propeller is the common cause for such rope kinking and snarling, but neither buoy had any propeller gashes.

"What's going on?" I exclaimed out of exasperation. Noticing the next trawl's end buoy was missing, I quickly began hauling that trawl with its remaining buoy showing. My hydraulic pot hauler began to labor, yanking up the last pot when suddenly the heterocercal tail of a nearly lifeless basking shark broke the water below the hauling block. I fastened down the trap line before bending over the railing and peering down into the water column. The entangled shark lay there vertically.

Somehow, this huge, harmless plankton feeder, which can grow to over thirty feet long, had first brushed the rope and then twisted to escape, eventually spinning itself and the buoy line right down to the trap to the point where it could no longer move and breathe. These sharks must keep moving to breathe. Basking sharks and sometimes whale sharks—also planktivorous and the largest fish of all—along with many of their carnivorous relatives, including mako and blue sharks, commonly invade coastal waters during August when surface water temperatures often hit seventy degrees Fahrenheit. Occasionally, the sight of the basking shark's dorsal fin breaking the water off of beaches erroneously prompt shark alerts, causing some bathers to needlessly flee the water. No doubt, this creature had freed itself after snarling and twisting the first buoy line.

Before cutting the eighteen-foot-long shark (the entire length of my skiff) free, I slowly towed it alongside my boat backwards to get a closer look at it. I could feel the deep-bodied gray and white shark sending weak muscular contractions along its body. After I cut it free, the shark sank slowly. I felt badly about the entanglement, hoping the creature would make it.

Then there was the case of nearly catching that record-size lobster which every lobsterman dreams about. The late Lane's Cove lobsterman Waino "Moe" Ahonen came alongside of me in his lobster boat in Folly Cove on a summer morning in 1965, wide-eyed and out of breath with a remarkable story to tell. He had just hauled up one of his thirty-six-inch-long wooden traps, which had a monster male lobster resting on the top of the trap with its claws even extending over the front. Ahonen figured that lobster weighed about forty pounds.

The lobsterman had cautiously winched the trap toward the surface, and he readied himself over the rail to grab one of the lobster's claws. To his dismay, the creature slowly slid off the trap, and Ahonen could only watch it sink slowly out of sight into the seventy-foot depths.

Four rare sightings, including three of living things and another of an inanimate object, make up more frozen memories. While steaming to my lobster gear in Ipswich Bay one torrid August dawn, an approximately five hundred-pound giant blue fin tuna suddenly rocketed itself twelve feet out of the water about five hundred feet ahead of me. Within seconds, the powerful fish slammed into the ocean's surface sideways before disappearing. I chuckled, "Is this a sub missile attack?"

Also, while hauling lobster traps near the Sandy Bay breakwater in May 1997, I thought I saw a horse's head silhouetted against the distant sky at the southern end of the breakwater. Tiers of pinned-together twenty thousand-pound-apiece "header" stones make up the breakwater's central high part, which shows during all tides. The tide had fully ebbed that morning.

A close inspection revealed a bewildered and tide-stranded nine-foot-long male gray horse head seal. In my presence, the mammal repeatedly drooped its characteristic horse-shaped head over the edge of the breakwater, thinking about escape routes. The seal appeared miffed as to where all the water had gone. He had about a ten-foot drop to the water. A quick look satisfied my curiosity; I did not want to further stress the seal. He was gone the next day.

"That bird holding something with its talons atop a tooth-shaped rock in the boulder field behind Pigeon Cove Harbor looks awfully large," I said to myself while returning to that port one fall morning of 1995. Lo and behold, an approximately four-foot-high immature bald eagle with coppery plumage feasted on the final remnants of a black-backed gull: its wing. Scattered feathers littered the eagle's impromptu dining area. That gull was probably the same one with a broken wing who had sought refuge in Pigeon Cove Harbor.

The eagle let me come to within twenty-five feet of it in my skiff. I'll never forget the piercing, wild look coming from its yellow eyes as it occasionally glanced at me while ripping at the wing's flesh with its razor-sharp curved beak. That eagle, soaring high in the sky above the shoreline, caught the attention of many Cape Ann birding enthusiasts for the next two weeks. Then the bird disappeared.

A bitter cold January morning around 1990 set the stage for the bizarre sighting of something inanimate. I had just left Pigeon Cove Harbor, guided by compass and sounding machine for the approximately mile-and-a-half-long journey across Sandy Bay to the Sandy Bay breakwater. A gentle northwest wind exercised in the five degree Fahrenheit air as a layer of dense sea smoke carpeted the ocean just high enough to block the visibility of yonder land. The arctic sea smoke's wisps frequently billowed up into the clear, sunlit sky. Any spray immediately froze on the rail.

About a third of the way across the Bay, the gray superstructure and smokestack of a huge vessel suddenly popped up about four hundred feet ahead of me. Sea smoke hid the rest of the vessel. Then it slowly disappeared in the sea smoke, heading north. "Was this a mirage? I thought you only saw those out in the desert!" I exclaimed, continuing on my course.

The afternoon's rising temperatures dissipated the sea smoke, revealing the morning mystery: an approximately two hundred- to three hundred-foot-long Canadian naval destroyer that had sought the sheltered Bay's ninety-foot depths that morning. The warship, on a training mission, stuck around Cape Ann waters for several more days before heading elsewhere.

Rare Personal Experiences

This Joe Sinagra drawing depicts the basking shark entanglement incident.

Just about every lobsterman gets stuck in some sort of nasty, heart-beating weather. One September squall in 1999 packing thunder, sizzling lightning, gale-force wind and torrential rain pawed with me at dawn off the Sandy Bay breakwater.

Earlier, the eastern horizon was clear as a bell, and the ocean's surface was glassy. Soon, a sooty-colored squall line rapidly bore down on me from the northwest. Suddenly a rush of cold air grayed up the ocean's surface. The sky darkened to pitch black, giving me just enough time to don my hooded raincoat before all hell broke loose. Huge raindrops cratered the ocean's surface as the gale-force northwest wind both churned the ocean into a three-to-four-foot-high feather-white chop and drove the torrential rain horizontally. Thunder and omnipresent lightning crackled and sizzled. I immediately shut off my VHF (very high frequency) radio and color depth sounder, badly affected by all of the air's electricity.

Rather than return to port, I decided to ride this storm out. "Ah, this will just be another quick-moving squall," I reassured myself.

Well, the quick-mover dragged on and on. All I could do was jog slowly into the blinding wind and rain and building waves. Despite a spray hood and windshield, I had to turn my head away from the elements much of the time. Some of the chops curled over my bobbing skiff's bow as the outboard's lower unit often came out of the water, belching exhaust coughs. The petrifying thunder and lightning kept coming. The air constantly sizzled with electricity all around me. I hoped that one of the potentially deadly lightning bolts wouldn't target my skiff. Luckily, after a very long forty minutes, the wind and rain abated, and the sky finally lightened.

Slowly, the powerful cell moved to the southeast, still making its might known by distant thunder and lightning. The ocean quickly moderated, and the daylight returned. The air was so fresh. "Good riddance," I said before resuming lobstering.

Lastly, there were five animate and inanimate rescue memories, including two incidents when I even had to rescue myself. The first scenario involved saving a lobster trap of mine that was "hung down solid" and also removing the cause of this predicament. After repeatedly failing to free the wooden trap by tying its line to the side of my boat and going around slowly in circles, I dove forty feet down on flat ledge about a quarter mile off the Lanesville shoreline with scuba gear in July of 1970 to do the job once and for all.

The visit to this wet, silent world revealed the trap was hung on a 2,500-pound anchor measuring 6 feet high and 9 feet long with an 8-foot-long stock. Three other ghost traps met my trap's fate, too. Clouds of curious cunners hovered around me as I cleared my trap and marked the anchor with a separate buoy and line.

The salvage occurred three days later, aided by a tow rope, a modified 250-gallon oil tank with a chain harness and a hole in its bottom, a spare air tank with a regulator and lobsterman friend Captain Winthrop "Bunt" Davis aboard his thirty-six-foot-long, diesel-powered lobster boat dragger *Carolyn Rene*. That size tank should be able to float two thousand pounds. I soon sank the tank, chained it to the front part of the anchor and displaced the water inside with air from the spare tank. I saw rusty water coming out of the hole in the tank as it slowly rose and its chain tightened. The oil tank wasn't quite buoyant enough to float the anchor; yet I could move the front part of the anchor up and down a little. I also fastened the inch and a quarter diameter nylon towline to the anchor's fluke. Next, Davis towed the tank and anchor alongside a wharf in a nearby cove where a wrecker yanked them out of the water onto its flatbed right to my front yard.

I donated the anchor to the City of Gloucester, now sitting in front of the Ralph B. O'Maley Middle School in Gloucester with a fresh coat of black paint. The anchor probably came from a two-masted schooner carrying coal that went ashore in that area during a northeast snowstorm around 1900.

The second rescue story was filled with irony: I had to kill one animal to rescue another. A distant bobbing object on the ocean's flat surface caught my attention this June morning of 1979 off Plum Cove beach. "It's probably just a seal's head," I initially reasoned. But the wobbling and bobbing continued, prompting a close look. I couldn't believe my eyes: there was a black-backed gull stuck in the mouth of a three-foot-long monkfish lying diagonally in the water. The rear end of the gull stuck out of the fish's mouth; the bird's tail fanned and its webbed feet struggled in the air.

Rare Personal Experiences

This Joe Sinagra drawing recreates the standoff between the huge monkfish and the black back gull. The gull was later rescued at the expense of the fish.

The sight of the gull swimming on the surface or its webbed feet movements must have caught the monkfish's attention. Eyeing a good meal, the stealthy fish next swam up and grabbed hold of the unsuspecting gull headfirst. Consequently, this snapping turtle of the sea and bird got stuck in a lethal standoff at the surface. Neither could the bird escape and fly off, nor could the fish swallow the large, buoyant bird and get back to the bottom where it belonged.

Another name for the monkfish or anglerfish is the goosefish, for the very reason that its diet often consists of waterfowl caught in the same manner that fish used to catch the gull. Gloucester gillnet fisherman, B.G. Brown, has netted monkfish over one thousand feet down along the continental slope off of Georges Bank with seabirds in their stomachs. This shows how far up these fish will rise to catch prey.

Consequently, I yanked both creatures onto my skiff's railing, and then pried the slimy fish's mouth open just far enough for an inward peek. The gull's head was lodged sideways in the fish's upper esophagus; one of its eyes stared up, giving the look of death. Its body, held fast by scores of inward-pointing teeth, had numerous puncture wounds.

"Should I play god?" I pondered, reasoning, "I'll save the gull since birds are more advanced animals than fishes." I then pithed the fish's brain and severed its jaw muscles, reaching into the mouth ever so slowly and freeing the bird. But not before the gull bit down on my right forefinger with a death grip. Consequently, I returned the traumatized bird to the water where it briefly remained motionless before finally releasing my smarting finger and then flying off.

I was out lobstering near Andrews Point along Cape Ann's northeast coast during the third rescue and retrieval incident one early May Saturday morning in 1982, struggling at times to stand upright in my thrashing boat. The shoreline and its outer ledges were white with the thundering surf of an offshore storm. Foam trails streaked the lumpy, chalky-green ocean's surface. A salty smell permeated the air.

The sight of nearby flashing lights from U.S. Coast Guard and Rockport Harbor Patrol vessels circling to the north prompted me to conclude, "Oh, no, another poor soul must have gotten swept off of the rocks." Around 6:00 a.m. the late Rockport Harbor Master Gene "Shorty" Lesch came alongside of me in the Rockport patrol boat asking, "Keep an eye out for a body."

The victim, a nineteen-year-old Boston College student, and two classmates had driven that Saturday from Boston to watch the sunrise at the water's edge. No doubt they had gotten too close to the edge, probably oblivious to the sea's dangers, only to be swept in by a series of rollers which most likely came from nowhere. The survivors attempted to rescue their friend, but the flowing tide quickly swept him away.

While hauling more traps in the vicinity, I homed in to two clues that helped me find the body. Foam trails mapped the tide's southward movement, also paralleling the stretch of shoreline there; and seagulls hovered above something in those foam trails. Sure enough, about three hundred feet off the coast in back of the Ralph Waldo Emerson Inn, the young man's body floated face down. An air bubble in the back of his navy blue sweatshirt kept him afloat, upright. I immediately radioed the authorities, standing by until they quickly arrived and retrieved the corpse. A medical team at the Addison Gilbert Hospital in Gloucester pronounced him dead at 10:57 a.m. after trying in vain for four hours to revive him. The ocean seems to claim one or two surf viewers or daredevils a year on Cape Ann.

I've also had to rescue myself from two serious situations. As I left Pigeon Cove Harbor to lobster on a summer day of 1988, my seventy horsepowered outboard suddenly stalled. Luckily, the gas hose had simply become disconnected at the motor. After reconnecting the hose and priming its bulb, the outboard fired right up.

My lobstering day went along smoothly until I was hand pushing one four-pot trawl off the railing. With three wire traps already out and sinking in the water, some of the remaining kinky ground line on the cockpit floor suddenly popped up and somehow wrapped around the palm of my gloved right hand. With the boat going a quarter speed ahead, I was immediately yanked aft on my knees and my entangled right arm was pulled out straight over the stern. I prevented myself from being yanked overboard by buttressing my right shoulder against the stern corner and its overlying A-frame, right alongside the outboard.

Rare Personal Experiences

I was trapped: the outboard had neither external controls nor a kill switch. There was no emergency knife at the stern to cut the line, and there was no one else nearby to hail down. The previous day I had dropped my regular knife overboard, usually kept amidships. The emergency knife acted as its temporary replacement. Luckily, my left hand was free.

The mounting strain of the three trailing traps not only cut the twist deeper into my palm, causing excruciating pain, but also increased the chance of my arm getting pulled out of its socket. I repeatedly attempted to gradually pull in the trailing line and traps with my right hand, hoping to reach over and grab the line with my left hand to take the strain off the entangled hand just long enough to free it. This failed time after time until my arm muscles gave out. Then a desperate solution popped into my mind. First, tie the end of that trawl's buoy line to the stern's A-frame and kick off my boots, and then push that last trap off the rail and jump over with it, hoping there would be just enough slack in the line to quickly get my hand free before being pulled below the surface. The tied buoy line would prevent the boat from taking off. Once free, I could swim back to the skiff.

But suddenly, thanks to the earlier gas line incident, I envisioned a simple solution staring me in the eyes: reach over to the other side of the outboard with my left hand and disconnect the gas hose.

In no time, the outboard simply ran out of gas, and the strain was taken out of the trailing traps. After about fifteen minutes of vigorously shaking my hand to regain full circulation, I resumed lobstering, very sore, but still able to do the job. I learned once again, keep calm, don't ever give up, and there's often a solution to the dilemma before your very eyes.

During the last self-rescue incident on June 9, 1976, I nearly experienced that Boston College student's fate. I began my lobstering day that early morning by hauling a string of wooden traps in the shallows off Andrews Cove, a treacherous, surf-prone rocky cove on the northern tip of Cape Ann. There were no other boats in the area. Despite no wind, an offshore storm's heavy swell meandered, making the shoreline white. The morning's astronomical low tide compounded matters. I was cautious about not going in too close—or so I thought. I had been shipwrecked one other time trying to save lobster traps in close during a northeast wind.

While working on one four-pot trawl about two hundred feet from the shore with my eighteen-foot-long skiff's stern pointing seaward, I suddenly turned around in complete horror as a twenty-foot-high cresting swell bore down upon me. This steep sea had grounded out on an underwater hump, made worse by the extremely low tide. Reflex told me to gun the outboard, swing the boat around, and head into the monster sea. There was no time; the swell surfboarded me toward the barnacle-covered shoreline at about a forty-five-degree angle. I held onto the steering wheel for dear life, only to have the energy-packed swell first spin the skiff broadside and then flip it with a wall of water.

The next thing I remember was a thud on the top of my head as the portside railing struck and pile-drove me down into the fifty-five-degree water. Luckily, my feet didn't ground out, but I came up into the bobbing skiff's water-filled cockpit. Lobster trap rope stored in the boat entangled my feet. Even worse, the grayish darkness, bubbles and turbulence there destroyed my sense of direction. "Which way is up? Or down? I have to breathe. I have to get out of here," I remember thinking. Soon, a faint inner voice emanating from the heart said, "This is the end." Being a strong swimmer, in good shape and a scuba diver who once had his facemask suddenly ripped off by a buoy line ninety feet down, I didn't panic; I looked for a solution instead.

I somehow got out, still weighed down by boots and oilskins. My head broke the lumpy surface about thirty feet from the shore and I parted company with the overturned skiff. I gulped down fresh air into my aching lungs; how good it tasted!

But unfortunately, my ordeal hadn't ended; I had to reach the shore and then get out of the turbulent ocean. My immediate concern was to stay afloat, assess the situation, then strike at a window of opportunity, and most of all—keep my cool and don't give up. I couldn't feel the cold water temperature at all. The boots were the first to go. One swell would carry me in, often to within an arm's length of the shore, only to have its backwash suck me right back out. This continued for a guesstimated twenty-five minutes. My movements became sluggish. No doubt, hypothermia was gaining ground. For a second time that inner voice repeated itself, "This is the end."

At last, a chance arose, and my left hand's fingertips death-gripped onto the sharp edge of a barnacle-covered rock. That swell's backwash tried its hardest to rip me free, but my stocking-covered feet soon touched down on solid ledge below, and the surge retreated, leaving me high and dry for the moment. With my adrenaline rushing, I quickly clambered up slippery rocks and headed for the nearest house about a mile away. An aged couple responded to my desperate door knocks. They gasped seeing me in their doorway, waterlogged and shocked with a goose egg on the top of my head. The Rockport police quickly transported me to the Addison Gilbert Hospital where emergency room staff observed and tested me. I refused the hospital's request to stay overnight. The blood test revealed that I had swallowed no water. Once home, I went right to bed, shaken and exhausted, only to be awoken several times in the night by nightmares. The nightmares lasted about two weeks.

In the meantime, word of the mishap quickly got out. Some fellow fishermen responded by salvaging all that was left of my fiberglass skiff: the bare hull. The seas and subsequent pounding on the shore had knocked off the outboard, hydraulics, spray hood and the hauling station. The boat was insured—declared a total loss by the insurer.

I couldn't wait to get back out lobstering. Shortly afterwards, other lobstermen let me borrow their boats to tend to my gear. Within a month, I was back in business with a new boat. I also gained a greater respect for the ocean as well as a new depth to life, learning that death by drowning wouldn't be a bad way to go; it would be over before you knew it.

IV
Lobster Tidbits

The American lobster, which can grow over three feet long and weigh over forty pounds, has survived the test of time so far, even withstanding the fishing pressure and the advanced gear of recent times. The lobster is an animal which, like most other living things, possesses a gene pool with unique characteristics, exhibits specific behaviors, occupies a special niche in the marine ecosystem and has a prime purpose to help balance marine nature and also to be food for others in the pyramid of life and to survive, grow and propagate its species.

Scientists and fishermen agree, "We know a lot about lobsters, but there's a lot we don't know about them." "If the lobsters ever get figured out, they will be in trouble," says Mark Ring, a veteran lobsterman out of Gloucester.

Chapter 4 crawls into this creature's classification, some external anatomy and common and not-so-common features, responses to environmental challenges, geographic range and habitats, migrating, mating, and molting behaviors, life cycle and diet and enemies.

Classification and Some Basic Anatomy

The American lobster belongs to the largest invertebrate phylum, Arthropoda, class Crustacea, order Decapoda, infraorder Astacidea, genus *Homarus americanus*. Unlike the other invertebrate phyla, the arthropods, including shrimps, crayfish, crabs, barnacles, insects, copepods, horseshoe crabs and other lobsters, inhabit the air, water, land and even the insides and outsides of other creatures as parasites. Making up approximately 80 percent of the animal species, the arthropods have specialized jointed appendages and segmented bodies covered by exoskeletons—adaptations which afford mobility and protection.

As described in Hickman's *Biology of the Invertebrates* "The decapods may be divided into the long-tailed members [shrimps, lobsters and crayfish] and the short-tailed members [crabs]. The long-tailed forms have a cylindric or laterally-compressed carapace, usually with a rostrum [which is the sharp tip of the head]."

Three major sections make up the lobster's body: the abdomen, thorax and head (cephalothorax); the latter two are covered by a shield-like carapace (a protective covering over its back and sides), which lobstermen often measure with a gauge to make sure it is at least the 3¼-inch-long minimal legal size. The abdomen, thorax and head are further made up of segments, each containing a pair of jointed appendages, including anterior feeler-like antennae and mouth parts, five pairs of mid-section walking legs (hence the name "decapod") and also five pairs of fan-like swimmerets on the underside of the tail. The appendages aid the lobster with grabbing, crushing and ripping apart food, feeling, swimming, walking, defending and mating. Lobsters also grip with their tails, especially in the corners of lobster traps. The larger seizer (or cutter) and crusher claws—the cheliped—are the first set of the lobster's mid-section walking legs. Random surveys of mine conducted while out lobstering have shown approximately 50 percent of the lobsters have right-handed crusher claws.

The infraorder Astacidea specifically consists of lobsters that have claws. The American lobster has two other such relatives who also live in the ocean: *Homarus gammarus* (the very similar-looking European lobster, dubbed the "blue lobster" because of its color) and *Homarinus capensis* (which grows to just under four inches in length and lives off of South Africa).

Males vs. Females

Adult lobsters exhibit sexual dimorphism. Five external features clearly differentiate a sexually mature male lobster from a female. First, the male's tail is more tapered than the female's uniformly-shaped, flattened one, specifically designed to hold and protect her fertilized eggs. Next, the male's crusher and cutter claws and body trunk are larger and stockier compared to the female's smaller and narrower counterparts. Also, the male's first pair of swimmerets—the gonopodia—is much longer and stiffer than the female's short, feathery versions. Lastly, the female's gonopore, where her eggs exit her body, is located between the bases of her third walking legs, while the male's is positioned between his fifth pair.

Geographic Range and Habitats, U.S. Zones and Lobster Groups

Homarus americanus, also dubbed the cold-water lobster by its purveyors, lives from Labrador down to North Carolina, from the low tide mark out to just beyond the continental slope, sometimes over 2,500-feet deep. During the summer and fall molting season, lobsters hiding amongst exposed rocks can be heard bubbling (breathing) at low tide. These creatures live on and in the muddy, sandy, gravelly, rocky, clay and even coral tree bottoms of submarine canyons, basins, ledges, banks, saltwater rivers, coves and

During my active diving days, I picked up some big lobsters like this 19 ½ pound male. This lobster was sitting right in the open forty feet down, alongside a muddy, rocky edge.

harbors. Lobsters prefer shelter to cover their vulnerable tail ends, and they often make shelters by digging out holes in the mud ("dishes") or crevices under or between rocks. These animals will also quickly make homes in discarded tires.

The Lobster Stocks

Most American lobster biologists, managers and fishermen have recognized three lobster (stock) zones off of the U.S. east coast: the Gulf of Maine, Georges Bank and south to the Carolinas (mainly offshore), and Cape Cod south to Long Island Sound (inshore); that is, until 2005. The lobster powers that be are now talking about breaking the lobster stock zones into Gulf of Maine, Georges Bank and Southern New England. Some mixing of these stocks occurs.

Also based largely on the lobster's size and degree of annual movement, many of the zones' stocks further consist of three groups: the short-distance migrators, the groundskeepers and the long-distance migrators. Groundskeepers, often dominant males, stay put most of the year, occupying the best homes, especially rocky crevices. This group makes up much of the winter and early spring catches in areas where most of the other lobsters have moved on.

Throughout the year many of the short-distance migrators travel between the shallows and their adjacent wintering grounds. These smaller lobsters, large enough to move and generally weighing below two pounds, not only make up the traditional Fourth of July and fall shedder runs, but also most of the annual lobster landings.

As one might expect, the long-distance migrators are the larger and stronger lobsters, commonly two pounds and over, especially the big "fan-tail" females who arrive in the shallows as the spring run seeking warm water to either hatch their eggs, molt and then mate, or in the case of many, spawn new eggs, becoming the "black or green eggers". These females, with their newly spawned tiny, dark eggs, dominate the inshore until the water temperature drops there, usually by late October. While many of the smaller eggers seem to stay inshore, the larger ones move offshore to seek warmer water so their eggs will continue to develop. Jan Robert Factor writes in his *Biology of the Lobster*, "During winter, lobster eggs become dormant." The arrival of big lobsters inshore weighing five pounds or greater usually signals the end of the spring run here.

Former offshore lobsterman and now shore captain Bro Coté of Milton, Massachusetts, has fished Georges Bank more than twenty-five years, lately aboard his seventy-five-foot-long Hyannis, Massachusetts, based *William Bowe*. Cote seasonally traps lobsters throughout the year from 20 fathoms down on top of the Bank all the way down to its slope's 280-fathom bottom depths. According to Coté, "There are two main lobster stocks out here—the migratory stock, which moves out of the deep in the spring all the way up to the top of the Bank, and the deep-water lobster which also moves up but stops often at the 100-fathom edge." The lobsterman has also found that the Georges Bank stock is reinforced with long-distance migrators moving down from the Maritimes and groundskeepers, especially the big males, "control a lot of the bottom," describes Coté.

Cold-Bloodedness

In sharp contrast to the warm-blooded birds and mammals whose body temperatures remain fairly constant, lobsters are cold-blooded animals whose annual activities, including feeding, moving, molting, mating and spawning, are largely governed by the surrounding water temperatures. "The water temperature and weather control lobstering," explained fifth-generation Nova Scotian lobsterman Jerry Himmelman. Generally, the warmer the water, the more active the lobster.

The lobster generally doesn't do too well both in the water and out of the water at temperature extremes. They are sluggish or dormant when the water temperature is too cold—usually below forty degrees Fahrenheit—and half-dead or sleepy when it's too warm—usually above seventy degrees Fahrenheit. In contrast, lobsters, especially new-shell ones, will die in the heat out of the water, or they will drop their claws and even freeze to death in the cold, since their body is mostly water.

Lobster Tidbits

Striped bass are one of the inshore lobster's greatest enemies today. These lobsters were in the stomach of this bass, which had just been filleted.

Water temperatures naturally fluctuate with the seasons, being the warmest from August through mid-October, usually in the high fifties to the high sixties off Cape Ann, and the coldest from January through March, sometimes dropping to thirty-five degrees Fahrenheit. In the coastal Cape Ann area, the majority of the lobsters remain active between forty degrees Fahrenheit and sixty-eight degrees Fahrenheit. Temperatures around thirty-five degrees Fahrenheit shock most of them into hibernation and others into near inactivity. They will still feed then if the baited trap is dropped practically right on top of them. This is why traps usually need to be set a week or so to catch lobsters during the winter. But I've had cases where a lobster will go right into a trap during the dead of winter.

There is no set date to the often-abrupt end to the lobstering season in the winter. I've seen the lobsters quit in December, most of the time by mid-January, and sometimes in February. In contrast, rising bottom water temperatures in the spring reactivate lobsters. A fifty-degree Fahrenheit water temperature (or better) seems to be the magic temperature that gets them moving in earnest. Cape Sable Island lobsterman Richard Nickerson has found lobsters begin to trap more off Nova Scotia once the water temperature rises to thirty-eight to thirty-nine degrees Fahrenheit.

The lobster's level of pre-winter built-up body energy reserves might explain the timing of either full or semi-hibernation. Calm autumns when lobsters can feed uninterrupted and build up energy reserves might cause them to hibernate early. In contrast, big fall storms like the October 1991 No Name storm and December 6–7, 2003—"classic storm '03"—probably interrupted their fall feeding and left the survivors hungry with no choice

Will this lobster feel pain as it is thrown into a pot of boiling water? Latest research has found that lobsters don't feel pain; they instead react to hot water by flipping and flapping. Anyone concerned about a lobster feeling pain should refrigerate the lobster before cooking it. The cold will put the lobster into a state of hibernation.

but to feed in the dead of winter. Lobster catches picked up the winter after the No Name storm, and many lobsters kept feeding during the winter of 2004 despite it being one of the coldest on record.

Wind direction can also temporarily affect the water temperature any time of the year, and thus lobster activity. Strong offshore winds, which blow warm surface water seaward and replace it with cold bottom water, often shock these creatures into brief inactivity, even in July and August, by rapidly dropping water temperatures by five to ten degrees Fahrenheit. In contrast, onshore winds, which around Cape Ann chiefly blow out of the northeast, do the opposite and warm up the water a few degrees. This frequently speeds up the lobsters' metabolism and makes them trap better, even in the winter.

People often ask, "Do lobsters feel pain, especially when dropped into boiling water?" They obviously feel something, since lobsters usually respond to that by flipping their tails and moving their claws. It's anybody's guess whether the lobster is reacting to pain or to a change in temperature. Although they are advanced invertebrates, the lobster's systems—including the central nervous system—are primitive compared to the more-evolved vertebrates. I think the lobsters flip-flopping in hot water is more a reaction to a change in water temperature and not pain. A 2004 study in Norway focusing on whether or not lobsters feel pain concluded, "lobsters don't feel pain. The movement is an escape mechanism and not a conscious response or indicator of pain."

In their Fish Empathy Project, the animal rights group, People for the Ethical Treatment of Animals (PETA), used this slogan "Being Boiled Hurts—Let Lobsters Live" to discourage people from killing lobsters.

I usually chill lobsters into inactivity by first placing them in the refrigerator for several hours before dropping them into a pot of boiling water. This way they do not know what hit them.

Lobster Movements—Where, How, When and Why

Although the ocean surface might appear still at times, that does not mean tremendous movement between currents and live action isn't going on below. Lobsters not only have legs, but also, like fish, have tails. Lobsters, like just about every wild animal, do their annual "thing" to grow, survive and perpetuate the species. They can and do move. Lobsters often travel in similar size and sex groups. Like migrating birds, the males seem to arrive first. Much of the life of the older lobster is spent on the go, seasonally finding optimal water temperatures and locales to molt, mate, spawn, hatch eggs and best survive often-turbulent winters. Generally these crustaceans gravitate to the warmer and safer shallows of the coast and tops of offshore banks and ledges during the early summer and fall. On the other hand, during the winter, many go off the edge and onto deeper, smooth bottoms (two hundred feet or greater) along the coast as well as those of inshore and offshore banks, ridges and ledges. This is the lobster's equivalent of heading south, like so many of the migratory birds. Most of the lobster's springs and falls are spent going to and from these depth extremes. Some of these movements are short-distance annual round trips; others are long-distance ones—sometimes hundreds of miles long. Lobsters know no borders.

Former long-time Georges Bank and outer-Gulf of Maine lobsterman and now also shore captain Charlie Raymond from Beverly, Massachusetts, owns the seventy-seven-foot-long *Michael & Kristen* out of Gloucester, Massachusetts. He summarizes one of the common Georges Bank's yearly lobster movements: "You see a run from the edge to Georges [the top] and then back down again." A crude lobster tagging study by Raymond and crew one spring, which began along the deep outer slope of Georges Bank, verified this. "One winter we marked some lobsters there by banding their knuckles [with rubber bands]. A couple of months later in the spring, we picked up several of those lobsters fifty to sixty miles away, further up on the edge of the Bank," says Raymond.

In the largely confined western Long Island Sound (at least before the 1999–2000 lobster die-off, to be briefly described later in this section), the lobster's annual movements were minimal. Here, they pretty much stayed put on the Sound's "sooty" mud bottom, residing in the forty-five-foot-deep shallows during the summer and in the one hundred-foot depths in the winter.

Prior to the historic 1999–2000 die-off, the Long Island Sound lobstermen used to "fish them so hard that we rarely saw any big lobsters—maybe one two-pounder a year. This was mainly a chix and pound-and-a-quarter fishery," says area lobsterman Joe Finke. Chix is an industry standard lobster grade for the smallest legal-sized lobster, which in the United States generally weighs a little over a pound.

Fascinating inshore and offshore lobster tagging studies by noted American and Canadian lobster biologists A.B. Stasko, H. Campbell, Jay S. Krouse, D.S. Pezzack, D.R. Duggan, Richard A. Cooper and Joseph R. Uzmann have gradually pieced together the complex inshore and offshore lobster movement puzzle, much of which has been seen firsthand by fishermen. With their southwestern Nova Scotia and Bay of Fundy mixed-size lobster tagging studies, Stasko and Campbell concluded that, "mature lobsters move farther than immature lobsters; in some areas mature females move farther and seasonally earlier than mature males; there was a seasonal shallow-deep migration of mature lobsters; and the long-distance movement of some mature lobsters allowed their mixing with other lobsters in the Gulf of Maine and adjoining continental shelf." One of the lobsters that Campbell and Stassko tagged and released at Alma, New Brunswick, was later recaptured offshore nearly five hundred miles away along the outer continental slope, south of Georges Bank. In 2000, Cape Ann inshore lobsterman Fred Hillier caught a lobster off of Rockport, Massachusetts, that was tagged at Fredericton, New Brunswick.

Also of great interest, Pezzack and Duggan found evidence in their Scotian Shelf-Brown's Bank Area study that, "long-distance return migrations and homing occur in lobsters. Some of the observed long-distance movements may represent a portion of a round trip migration." The inshore Gulf of Maine lobstermen regularly experience one such long-distance return migration involving the large-size brood stock including egg bearing and V-notch females. Many of these lobsters seem to move down the coast from either Maine or Canada to off of Cape Ann, traveling to the tip of Cape Cod, most likely via Stellwagen Bank, and then curve back up the coast. I call this the brood stock gyre. These females appear to be nearly always on the go. Fred Hillier reports "rarely if ever catching the same egg-bearing females in my traps."

Gloucester lobsterman Ron Hemeon fishes amongst this gyre off of Cape Ann. At times 90 percent of his catch is made up of females. With the passing of the zero tolerance law in 2003, he has had to often return 50 percent of his catch; in 2004 he figures that number increased to 80 percent.

In contrast, many smaller eggers seem to largely stay put inshore, warm water or not. It's not uncommon for inshore lobstermen to get two to three of these per haul during the winter.

Although awkward and weighed down by their heavy claws out of the water, lobsters can get up on their legs and easily walk underwater forwards, backwards and side to side with great agility—almost gliding. Sometimes these animals trek great distances over the ocean's often irregularly-shaped bottom, frequently wearing white the inner bottom edges of their large claws, then dubbed "scruffy" or "traveler's" claws by lobstermen. Scientist J.R. Uzmann estimates in *Biology of the Lobster* that some lobsters can move "four to five nautical miles per day."

Two veteran lobstermen give their insights as to what possible strategy some of the Gulf of Maine lobsters use to move about there. Cape Ann lobsterman Dennis O'Connell of Rockport, who traps lobsters ten to twenty miles offshore there during the fall and winter amidst the Gulf's basins, ridges, ledges and their edges. Aboard O'Connell's approximately forty-foot-long *Lady Elaine* he has found that, "Generally these lobsters [on the move] don't seem to climb steep edges; instead they pass through gullies and often follow edges."

Retired offshore lobsterman Stevie Robbins II of Little Deer Isle, Maine, who has fished the outer Gulf of Maine and Georges Bank aboard his former forty-four-foot-long *Shirley Freeman* and later fifty-four-foot *Stacie Vea*, concluded, "Lobsters seem to go the easiest path, usually following trenches and edges." But, he still vividly remembers one November lobster catch on the very steep northern edge of Georges Bank, "where the bottom came right up from 123 to 88 fathoms. The first three traps [of a 44-pot trawl] starting down each averaged 51 pounds of lobsters." Were these lobsters moving down the steep edge?

But, in the 1960s an old-time Lane's Cove lobsterman told me of seeing a school of lobsters swim backwards just below the surface about three miles offshore in Ipswich Bay while he was hauling tub trawls one spring day. Could these lobsters, especially the smaller ones, swim backwards off of the bottom by flip-flapping their tails with their claws extended? Isle au Haute, Maine, lobsterman Dick Bridges observed such schooling firsthand in the harbor there during the summer of 1996 as he cleared rope out of his boat's propeller. "About six half-to-three-quarters-of-a-pound apiece lobsters were schooling just below the surface," Bridges recalled. Frequently while scuba diving, I have witnessed smaller lobsters taking off from the bottom, flip-flapping backwards short distances to escape enemies.

I further wonder, could some of these lobsters also be hooking rides with currents like the one I experienced during a 1999 summer diving excursion to retrieve lost lobster gear along the Sandy Bay breakwater's seventy-foot-deep-edge? This strong and cold bottom current whisked me along with pieces of seaweed. From one of his coastal Maine lobster tagging studies J.S. Krouse concluded, "all distance migrants (greater than 20 nautical miles) appear to travel in the direction of the prevailing south to south-westerly coastal currents."

Responses to Environmental Challenges

A lobster's survival often depends upon quickly and properly responding to environmental challenges, especially big storms, low salinity and warm water. Occasional man-made and natural disasters have caught them by surprise, too.

Like some pets, lobsters, at least the inshore ones, are petrified by thunder and lightning. They hide in their homes from the bright flashes and piercing rumblings even hours after the storm has passed. Because of this, hauling traps that have just been set for one night the day after a thunderstorm has repeatedly proved to me to be a waste of time.

My experiences have also shown these arthropods will flee low salinities caused by heavy rains (at least around Cape Ann). Unlike eels, mummichogs (minnows) and salmon, lobsters cannot adjust to changes in salinity, which are measured in parts per thousand (PPT). Lobsters like salinities around thirty-two PPT. Too low salinity will cause lobsters to bloat and, even worse, explode and die.

I remember one spring deluge that turned Ipswich Bay's seawater brown and created strong currents there. Wave action also mixed the storm's ten-inch rainfall down deep in the water. Excellent lobster and flounder catches suddenly plummeted, returning to normal a week later once the water had cleared up. No doubt, those lobsters and fish evacuated out to deeper water where the effects of the storm weren't felt.

Old-timers used to say that the first big snowstorm would end that season's lobstering. Although many lobstermen consider this to be an old wives' tale, I have found this to be true in a few cases. The snow might suddenly lower the water temperature, shocking the lobsters into hibernation. When this occurs, the lobsters out in the deeper water become inactive about a week later than the ones in shallow water.

Quite often, these animals will flee warm water above seventy degrees Fahrenheit, at least off Cape Ann, and sometimes even in January along the continental slope off Georges Bank, after a warm Gulf Stream eddy has meandered there. Those exact conditions occurred during one offshore lobster-dragging trip in January of 1972 aboard the former Gloucester-based dragger *Judith Lee Rose* on Oceanographer's Canyon. The eddy not only created eerie sea smoke on the surface while being overridden by bitter cold Arctic air, but also drove away lobsters 1,200 feet down. The dragger was forced to steam away to colder water grounds to get back onto the lobsters. The eddy water, pumped aboard by the vessel's wash-down hose pump, was warm enough to swim in.

Inshore warm water, which has occurred in July but is more common during August off Cape Ann, traditionally drives lobsters (mainly the weaker new-shellers) from the heated shallows to the colder depths, especially off the edge of hard bottom. Already stressed by molting, the warm water further speeds up their metabolisms and stresses them more. The few lobsters that stay inside for some reason sometimes come up in the traps "sleepy" or limp and half-dead.

The warm water makes the lobsters susceptible to other diseases, especially shell disease. Warm water is also low in oxygen. Strangely enough, cold water, which is high in dissolved oxygen, also lowers the lobster's immune system, and like the warm water, also makes them more vulnerable to disease.

Lobsters seem to have a storm-fleeing instinct of leaving, often en masse, the potentially dangerous hard bottom shallows for the deeper and safer smooth bottoms. While scuba

Lobster Tidbits

Schools of codfish like these depicted in this Joe Sinagra drawing also eat huge numbers of lobsters, especially when they are on the move on smooth bottom.

diving, Maine Department of Marine Resources chief lobster scientist Carl Wilson has tricked lobsters out of crevices by banging two rocks together to recreate the sound of rocks being tossed about by a storm. Although the hard bottom, ledges, boulders, cobbles and even grout piles offer them the best shelter from predators, storm action can bang them up and, even worse, kill them there, especially new-shell lobsters.

Lobsters, who have gotten trapped in lobster pots before the storms, often come up dead in traps after storms, sometimes ripped to pieces by the turbulence, looking "like they have just come out of a cement mixer," according to inshore lobsterman Al Olson of Rockport, Massachusetts. Apparently, most animals can sense an impending storm or natural disaster. The majority of animals along coastal Southeast Asia had fled the lowlands before the tsunami struck in December of 2004.

The first big storm of late summer or early fall is famous for shaking up the lobsters and getting them to move en masse off the edge, often into the muddy basins of bays. Runs of these lobsters are mostly new-shellers. Once the ocean has calmed down, area lobstermen will usually "hit or slaughter" these travel-hungry lobsters, whose metabolisms have also been sped up by the storm's water-warming effects, getting huge catches in their traps at the lobsters' refuge sites.

Nowadays, such runs are shorter than years ago, now lasting about three to four days, compared to a week or longer before. This is probably a case of more fishing effort catching up the lobster pie quicker than before. But, in other situations, lobsters quickly move on. The catches begin with the healthy, larger lobsters and end with the "cripples," "culls" (one-clawed lobsters) and "ministers" or "bullets" (lobsters with no claws).

Storms can also cause soft-shell lobsters to get up and move and trap sooner than they normally would. Hence, a run of poor-quality lobsters will often hit the market and cause a drop in price.

If only one storm strikes, most of these lobsters will only move off so far, sometimes just hundreds of feet, and then return to the hard bottom. Successive storms, common in the fall, will cause them to keep moving further and deeper—sometimes right out to their three hundred- to four hundred-foot-deep, five- to ten-mile offshore wintering grounds.

After seeing a Jacques Cousteau TV special years ago on how the season's first big storm off the British Honduras caused the spiny lobsters there to march off the coral bottom and onto the nearby deeper smooth sand bottom, head to tail, in columns, I'm willing to bet the American lobster marches the same way.

One calm, balmy May day of 1965 gave me a good college of hard knocks education on the lobster's storm instinct. I was fishing a string of wooden lath traps—singles—in close along a notorious surf-prone stretch of rocky coast off Halibut Point (northern Cape Ann). Traps that day suddenly came up filled with lobsters. I couldn't figure out why, so I reset the traps right back in the shallow water. The lobsters knew, maybe feeling barometric changes. That day was a "breather" for a powerful three-day old-fashioned northeaster, which later washed ashore my entire seventy-five-trap string. Those trapped lobsters fed and trapped while trying to move to deeper water.

Lobsters occasionally misjudge storms and do not go out far enough, like many did during the near-hurricane-force Blizzard of '78 and the 1991 No Name storms, whose turbulences went down deep. Thousands of dead and dying lobsters and fish, even deeper-water cusk and redfish, got washed ashore on Cape Ann beaches, frequently mixed in with mounds of seaweed during those northeasters.

Man-made disasters, like oil spills and pesticide run-offs, can also catch lobsters off guard and kill them before they have a chance to escape. Two such disasters are described in the final chapter.

Molting and Mating

Molting and mating go hand in hand, and they usually occur in the summer and fall in shallow water when food is also plentiful. Offshore lobsters also move to the tops of banks and ledges to molt and mate.

To grow, a lobster molts or sheds its old shell, including the inner lining of its stomach and gills, growing about 15 percent in length and up to 50 percent in weight each time. During the early stages of molting or ecdysis, the lobster goes into hiding under various covers, especially rocky crevices and in lobster-made burrows in the muddy bottoms and bankings of harbors and salt water rivers. Most lobsters cease feeding and will not trap at this point. By then, the underside of its old shell has already dissolved away, and a rubbery, soft replacement has been laid down beneath the old shell.

During molting, the lobster first swells up with water and its carapace splits down the middle. Lobstermen call these lobsters "split backs". The translucent shell on the underside of the lobster's tail also becomes cloudy during molting. The carapace is the first to split apart and next, the lobster will actually pull out of its entire old shell. This stage takes less than thirty minutes. The resulting jelly-like lobster, or "mush ball," is very vulnerable. Remarkably, these lobsters or their nearby mates will consume the completely intact old shells for key ingredients to help build their new shells, explaining why old shells rarely wash ashore.

Once it has molted, the lobster will usually hide for another two to three weeks until its shell has hardened. Most of these creatures will not trap during the approximately six- to eight-week-long pre- and post-molting periods; although, a few do so at the last minute, being caught as either bloated-to-the-touch "split-backs" or as "mush balls" who have molted in the trap. Many of these "mush balls" get eaten up by their trap roommates. Most lobstermen return these soft lobsters to the ocean, since they have no economic value. Their claws won't even take a rubber band. Most lobstermen go by the rule, "If you can't band them, you can't land them." Incidentally, the rubber bands are used to keep the claws closed and prevent lobsters from biting one another. The bands, in contrast to the wooden pegs that were used years ago, don't puncture the lobster and weaken it.

No molting time is ever etched in stone, and the molting time can be area-specific. I've caught new-shell lobsters every month of the year inshore. I especially remember one warm January when new-shell, small males were plentiful in ten to twenty feet of water. Sizeable runs of new-shellers are now often caught on the deep muddy bottoms between November and January miles off from the shore.

The growing trend of more year-round lobstering further mid-shore and offshore during the cold-water months could cause those lobsters there to molt, too. The same lobstermen are now feeding lobsters, including many of the undersize lobsters just one molt away from legal size, which are either tossed back or leave the traps via escape vents, at a time when food is not normally plentiful. This man-made food abundance could well stimulate late-season molting, as wild creatures tend to multiply and grow faster during times of plenty. The bait scraps that settle to the bottom also feed lobsters in the area.

The younger the lobster, the more it molts. Scientists estimate that many one-year-olds have already molted ten times, and minimal legal size lobsters (whose carapace lengths are at least three and a quarter inches) have shedded approximately twenty times. The latter lobsters typically molt once or twice a year, making up the traditional Fourth of July and fall shedder runs.

But not all lobsters molt every year; lobstermen dub these the "skip molts". Egg-carrying females and lobsters in general who get caught in cold-water pockets and remain hibernating there for long periods fall into this category. The skip molts might shed their shells every two to five years. Some of these creatures, further described by the lobstermen as "Barnacle Bills," have large barnacles growing on their shells. The rock-hard shells of many of these skip molts require a sledgehammer to crack them. One Massachusetts lobster dealer who has cooked such lobsters says some of their shells around the claws and knuckles can be approximately one-quarter inch thick.

Lobsters inhabiting Gulf Stream-heated and climatically warm waters, including the shallows from Cape Cod south to Long Island Sound, generally grow faster than those living in the colder Gulf of Maine and Bay of Fundy locales. The lobsters in the cold waters usually need seven to eight years to reach legal minimal size, while those in warmer waters will do so in a year or two less. Also, male lobsters usually grow faster than females especially, as noted in *Biology of the Lobster*, "after sexual maturity, because the female must divert energy to egg production and delay molting to brood her eggs." People always want to know how old a twenty-pound lobster is. Who knows? One can only guess. It must be at least fifty years old.

Mating

A section in *Biology of the Lobster* further describes before and after mating behaviors, which usually get underway in the late spring when rising water temperatures trigger lobsters to move. The males seem to be the first to move.

> *Males establish themselves in shelters and await the visit of pre-molt females. The dominant male in the area…is usually the largest and is the one selected by the pre-molt female. During pair bonding, as the female enters his shelter, he fans the water with his pleopods [swimmerets], and the pair engage in a "boxing" ritual using their chelipeds [large claws]…One hour before she molts, the female engages in "knighting" behavior in which she faces the male and positions her claws on top of him. Knighting may signal the male the molting is imminent.*

The female will remain under the male's protection for about a week after the molting and mating, at which time he will also eat most of her old shell.

Mating activities usually occur right after the female has molted when she is clad only in her new soft shell. The two then couple with the hard-shell male on top of the upside-down female. The male, aided by his spine-like gonopods, next transfers his sperm, contained in sac-like spermatophores, to the female's seminal receptacle, located on the undersides of her main body between the walking legs, where it is stored for future use. The actual fertilization is delayed. This is described in the soon-to-follow life cycle description.

Dennis O'Connell is one of the few lobstermen to witness lobster mating at sea. This happened by chance one July morning in his boat's lobster tank. O'Connell had already put his first trap's catch of two regular-sized lobsters, one a male and the other a female whose claw shell was just hard enough to hold bands, into the water-filled tank. At first, "I thought the soft-shell lobster had died, since she was lying on her back with her claws extended. Next, the other lobster suddenly mounted that one, and I figured out they were mating. The mating only lasted about a minute before both got up and began moving around," recalled O'Connell.

On Georges Bank during the summer, Bro Coté has also noted that trapped "split back" females attract big males. These females act as good baits. They probably give off pheromones or biological messages. Coté has also noticed, "females will both go into and come out of their sheds first. The males do so about a month later." Brad Chase, a Massachusetts State biologist who has made some interesting naturalistic aquarium observations of mating lobsters, has a reasonable explanation of why this happens. "The males are very aggressive towards one another while competing for the [soft-shell] females. If a male molted then, the other males would quickly gang up and devour him." He also observed a male lobster devouring his mate's just-molted old shell.

Life Cycle

The lobster's amazing life usually begins about a year after mating and sometimes even longer down the road when its mother spawns thousands of black or dark green-colored eggs or oocytes, which have been ripening internally in the female's ovaries. From there the eggs take an interesting route, first traveling externally via her gonopore and then passing over the seminal receptacle where the eggs come in contact with the stored viable sperm, and fertilization occurs—thus the new lobster's life begins. The fertilized eggs finally move on to the undersides of her tail section ("the brood chamber") for attachment to her pleopods and hair-like setae there. This egg extrusion and fertilization usually takes place in August and September. By then the female has already carried these eggs in her ovaries for about nine months. The ovaries are the orange-colored mass that can be seen in some cooked lobsters' tail and body sections.

But, eggs don't always get fertilized during spawning, and these appear as empty orange-red eggs on the underside of the tail. These unfertilized eggs will eventually drop off. "Not all red eggs are dead. If there's any evidence of development, like eyes or a little lobster inside, the eggs have been fertilized, and they are all right," explains Diane Cowan, PhD, the senior scientist for The Lobster Conservancy in Friendship, Maine. The increasing numbers of such females with red eggs here could be explained by the inner Gulf of Maine's zero tolerance V-notch policy, which limits intake of female lobsters and therefore leaves an inadequate amount of males to fertilize these females. Ron Hemeon, a Gloucester mid-shore lobsterman, often catches up to one hundred of these lobsters during some hauls in the fall and early winter. Females comprise about 95 percent of his catch during this time. In contrast, close to the land, I might get two or three of these females a year.

While carrying the slowly developing eggs externally, the ovigerous female or "egger" becomes very protective, thinking nothing of crushing to pieces other smaller lobsters and even other "eggers" who might cross her path. "These lobsters are not very good neighbors," says Susan Jones, editor of the *Commercial Fisheries News*. While catching these eggers in the traps and before releasing them, lobstermen will often band shut their crusher claws to prevent these females from crushing and killing trapped lobsters in the future. Lobstermen will often put a band around one of the knuckles of a V-notched lobster's claws for quick identification. *Biology of the Lobster* reports on the habits of the mothers, writing that they

> *routinely aerate the eggs and clean them of fouling organisms. Fanning and grooming with the tips of the walking legs is critical to the health and successful incubation of the eggs…*
>
> *To survive the ten to twelve months of normal incubation in the wild, the embryo must have sufficient organic reserves, be securely attached to the pleopods, and remain free from disease, predation, and exposure to water of poor quality…*
>
> *Homarus americanus females normally lose 30 to 50 percent of a clutch during the long brooding interval.*

But fortunately a good percentage does survive. Typically in June and July (about nine months later), these now-swollen, pale brownish-gray ripe eggs will gradually hatch within several days from their mother's brood chamber into crude baby lobsters resembling mosquito larvae. Once hatched, the larvae become planktonic.

Here in the water column, they undergo four molts during a three- to ten-week period, at which time they grow more typical lobster appendages before settling to the bottom as mini-lobsters less than an inch long, with all the adult features. This brief planktonic existence is the most dangerous period during the lobster's life; many scientists believe that less than .01 percent of an egg hatching ever makes it to the bottom.

On the bottom the very vulnerable shelter-dependent lobster passes through juvenile and then more mobile adolescent phases before becoming an adult, characterized by, "the onset of functional sexual maturity," reports *Biology of the Lobster*. Generally, lobsters reach sexual maturity between five to eight years of age when they can weigh from a half-pound to slightly over one pound. Carl Wilson, the Chief Lobster Biologist for the Maine Department of Marine Resources, has found "that 60 percent of the lobsters in a natural population generally die before they reach the minimal gauge size [83 millimeters or with a 3¼-inch carapace length or a weight of about 1 ⅛ pounds], and 80 percent die before they reach the maximum gauge size [123 millimeters with a carapace length of 5 inches and a weight of about 4 pounds] which are generally over 15 years old."

1. Millard Crowley of Beals, Maine, was the state's longest lobster license holder. As of 1999, he had held a license for eighty-six years. Crowley passed away in 2002 at the age of one hundred.

2. Aspiring lobsterman Colby Polisson of Gloucester hauls single traps with his grandfather, Jason Polisson. Colby was ten years old when the photo was taken.

3. A lobsterman in Ipswich Bay checks his traps under a fall sky peppered by migrating surf scoters, a common harbinger for that season.

4. Ipswich Bay lobsterman Dave Eastman checks his traps during an approaching squall—very common during the summer.

5. Arctic sea smoke often occurs in January.

6. Rainbows are often the finale to summertime squalls, as shown by this Pigeon Cove Harbor shot.

7. Summertime dawns can be very dramatic, like this Salvages sunrise when the sky and ocean surface mimic one another with warm colors. The Salvages are rocky outcroppings several miles off of Rockport, Massachusetts.

8. A shot of a male lobster. His large, bulky crusher and seizer claws and tapered tail help identify his sex right away.

9. A shot of a female lobster. Notice how much smaller her crusher and seizer claws are in relationship to the rest of her body and also how much more flattened and less tapered her tail is compared to the male's. The wider tail is designed to protect her eggs, which she carries on the underside of her tail.

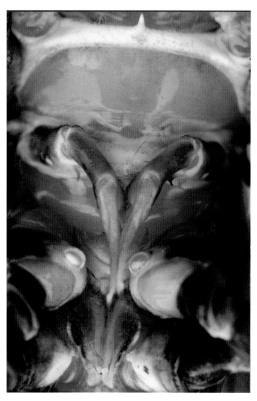

10. Shown here up close is the male's first pair of abdominal appendages, the gonopodia, which are long and stiff. One can quickly tell the lobster's sex by looking at these.

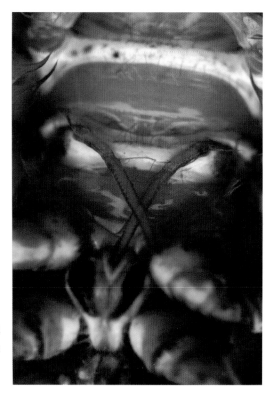

11. Here is the female's first set of abdominal appendages, which are soft and feathery and much smaller than the male's.

12. This female's dark-colored eggs were just extruded to the underside of her tail, where they will ripen for about the next nine months. Female lobsters usually extrude their eggs in the summer when the water is warm. Lobstermen refer to these lobsters as the "black or green eggers" because of the color of their eggs.

13. These female eggs are swollen and light-colored, signaling that they are ready to hatch into larvae.

14. Female lobsters sometimes carry reddish-colored eggs, which signal either that the eggs never got fertilized during extrusion and are dead, or they are just colored that way, and the eggs are all right. A close inspection sometimes reveals sets of eyes in the living red eggs.

15. This baby lobster is just weeks old. A dime can practically cover it. This little one dropped out of one of my traps and was spotted on the deck of my boat.

16. Lobsters spend much of their lives on the move, and when threatened, in this defensive mode.

17. Besides normal colors, lobsters also come in unusual colors including black, red, yellow, blue, white, spotted and even bi-colored like this one. Bi-colored lobsters and albinos are the rarest of the odd colored lobsters. Courtesy of Elaine Jones from Maine's Department of Marine Resources.

18. Lobsters can grow into giants. Here, the late fisherman Gus Doyle holds up a twenty-eight-pound male lobster that was dragged up in a net from 1,200 feet down. Fishermen refer to the big male lobsters as "horse lobsters."

19. Gloucester lobsterman Frank Davis displays a blue lobster that came up in one of his traps.

20. Shown here is a "split back" lobster, which is ready to shed its shell. The back of the carapace is the first part of the shell to break apart during the molting process. Most female lobsters mate right after they have shed their shells. Many lobstermen believe split backs like this one give off pheromones to attract males.

21. Ocean catfish, or wolf fish, like this one often go into lobster traps just to feed on the lobsters. These fish use their front incisor teeth to pry off mussels from rocks and their inner molars to crush them as well as lobster shells. These catfish can grow to be five feet long.

22. Stressed lobsters sometimes become afflicted by shell disease. Microbes, which feed on the shell, can make the lobster's exoskeleton resemble corroded metal.

23. Regularly colored lobsters have just about every color in the visible light spectrum. These claws show some of those colors.

24. Many lobstermen begin their day's work just after sunrise when the air is often cool and the ocean calm.

25. Rockport, Massachusetts, lobsterman Robert Hale, assisted by stern man Richie Adams, works traps aboard the *Krystal Marie*. Hale, like many lobstermen, fishes his traps in strings or trawls.

26. Northeast storms sometimes snarl up traps into bunches or blossoms. This is what the bunches look like at the surface.

27. Lobsters really taste good after steaming them right along the seashore. My family and I often have summertime cookouts at the seashore, where we first gather driftwood and then build a fire in a rocky crevice to steam lobsters in a pot.

28. Here's the finished product ready to serve, including steamed clams and hot butter.

Many lobstermen and scientists feel that upping the lobster gauge would give more females the chance to spawn at least once before being trapped.

Mainly depending on size, several molting, mating and spawning scenarios often occur for the adult females. According to this description from *Biology of the Lobster*, the smaller mature females' "reproductive cycle…typically takes two years. The female molts and mates one summer, spawns the following summer, and carries the eggs on her pleopods [swimmerets] until the third summer when they hatch. One to three months after the eggs hatch, the female molts again, mates to replenish her sperm supply, and the cycle is repeated."

On the other hand, many of the larger females, commonly weighing over two pounds each, skip the molting and mating between spawns. "Lobsters are usually able to fertilize two broods with sperm from a single insemination," as described in *Biology of the Lobster*. These lobsters are critical to the industry's future, since besides spawning more times between molts and matings than the smaller females, they also "produce relatively more eggs than the smaller ones." Wilson once counted 8,400 eggs on a lobster at minimal size and 41,000 eggs for a female at the 5-inch (around 5-pound) maximum size. Grand Manan Island lobsterman Lawrence Cook, who works Lobster Fishing Area 36, believes "the three- to six-pound females are the best breeders. The best breeder is not one found in the Old Folks' Home, nor is it the youngster." Many fishing areas don't land over-sized male and female lobsters with five-inch carapace lengths or greater and V-notched females to preserve the brood stock. Many fishermen and regulators look at these as an "insurance policy."

According to Diane Cowan,

> *The results from The Lobster Conservancy's sonar-tagging project reveal that large brood stock females travel significantly greater distances than smaller egg-bearing lobsters. Temperature data loggers attached to egg-bearing lobsters allowed for comparisons of water temperatures experienced during brooding…*
>
> *Large lobsters (greater than one molt above maximum legal size) traveled distances up to approximately 130 nautical miles (240 kilometers). Long-distance travelers released their larvae far from their spawning grounds. Small lobsters (within one molt of minimum legal size) stayed in the immediate vicinity and hatched their eggs locally. Hatching eggs locally and regionally affords many potential advantages to the lobster population…*
>
> *Spreading offspring over a broad spatial area increases genetic diversity. Genetic diversity is known to strengthen populations by producing individuals that are more robust to environmental changes…*
>
> *Dispersing larvae over great distances may also be important for reseeding areas where stocks have fallen off. For example, if a local lobster stock is decimated the area can be reseeded naturally as long as large brood stock lobsters are available and migratory routes are not interfered with.*

Special Lobster Traits and Oddities

Just as lobsters possess the unusual trait of being able to self-amputate appendages—most commonly their large claws—they can regenerate them, too. I've watched lobsters, stressed by either my handling of them or cold temperatures, drop whole claws in an eye blink. They don't seem to experience any pain in doing so. The break, which occurs in the joint close to the main body, quickly heals to prevent bleeding to death. While transporting lobsters in below-freezing temperatures from the boat to the market, I'll put them into buckets and carry these in the heated cab of my truck.

During regeneration, a red rubbery bud first emerges from the lost appendage's remaining stub. This bud gradually takes on the shape of the part-to-be, ranging from antennae to walking legs to big claws. Through repeated molts, the regenerated limb eventually grows to the size it should be. A small number of lobsters with such appendages are part of the lobsterman's everyday catches.

Unusual colors are one of this animal's oddities that catch immediate attention. Gino Mortillaro, manager of Mortillaro Lobster LLC in Gloucester, calls these "funky-colored lobsters." Lobster color "results primarily from the presence of pigments known as carotenoids in the tissues and shell," describes Dorothy E. Bliss in her *Shrimp, Lobsters and Crabs: Their Fascinating Life Story*. Since the lobster's biochemistry cannot manufacture carotenoids directly like plants, the arthropods instead get these indirectly by eating plants and other animals like sea urchins, herring and cod who have already done so somewhere along the food chain.

Most lobsters share a common dark olive top, which gradually fades with speckles and blotches all the way down to the light undersides, often pumpkin orange. The colors help the animals blend in with their surroundings. The lobster's antennae and the tips of its rostrum, claws and smaller walking legs are red. A close inspection reveals that many lobsters, especially the bright-colored new-shellers, contain all the colors of the visible spectrum.

Diane Cowan, who also writes "The Lobster Doc" column for *Commercial Fisheries News*, adds

> *Coloration in lobsters has many and varied explanations. However, primarily, it is genetically determined—much like hair and eye color in humans. The calicoes [lobsters with brown spots against yellow backgrounds] and those with white, yellow, and purple are born that way, and they stay that way. However, some of the blues, browns, and greenish, reddish and black tones can be influenced by things like diet, sunlight, and bottom type.*

Besides the calicoes, reds, greens, yellows, whites, blues and blacks, lobsters also come in rusty browns and even in very rare bi-color combinations such as red and green or red and white. Most lobstermen, including myself, catch a half-dozen or so of these oddities during their careers, with the blues and calicoes being the most common.

"Mud lobsters," which have dwelled in the muddy, cold depths where little, if any, sunlight penetrates for long periods, come up in traps with blackish-brown shell tops and whitish-brown undersides. Lobsters living amongst the rusted wreckage of an old steel ship off of Rockport have rusty-colored shells.

Several years ago, a lobsterman friend stored a blue lobster in a hanging crate off of his boat, feeding the crustacean herring for several weeks. Lo and behold, the lobster gradually returned to a normal color. No doubt, this animal's previous diet lacked the necessary carotenoids stored in the herring.

Claw and sex organ oddities also occasionally happen in lobsters. Some of these creatures are born with two of the same type of large claws, and in some cases, with claw deformities, including two or more of the small hinges or dactylopods per large claw.

A few lobster gynandromorphs have both male and female sex organs. Such rare lobsters have one stiff male gonopod and one female soft and feathery version as the sex organ pair, and the presence of external eggs only on one side of its abdomen. The female side of the lobster is usually on the right-hand side. The Pigeon Cove Harbor father and son lobstering team of Bob Morris Jr. and Bob Morris III trapped a two-pound gynandromorph lobster, aboard their lobster boat *Spirit* off Rockport, Massachusetts, on January 17, 2006. I caught one of these rarities about four years ago.

Prey and Enemies

Lobsters hunt and scavenge all hours of the day, either at or away from their shelters, sometimes pursuing interesting smells, especially lobster baits, carried by the currents. They are good ambush predators, too. Just about every neighboring creature small enough to kill, ranging from invertebrates (mussels, clams, sea urchins, crabs, planktonic copepods, fellow lobsters, etc.) to fishes (cunners, flounders, small pollock, cod, hake, etc.) to even "plant material," according to *Biology of the Lobster*, and just about every dead thing with nutritional value that settles on the bottom, including lobster baits and seals, make up the lobster's diet. Larval stage lobsters filter feed plankton out of the water.

The lobster's list of enemies includes microbes, invertebrates (chiefly other lobsters and crabs), fish (most notably herring, menhaden, cod, sea ravens, wolf fish, striped bass, ocean pout, skates and dogfish), sea gulls and cormorants, seals, whales and sharks, man and Mother Nature. The younger the lobster, the more enemies it has. Lobsters are noticeably cannibalistic during the molting season when new-shellers crave their fellow lobsters' shell components for the growth and hardening of their new shells. Extremely protective egg-bearing females often crunch up other potential lobster threats, too. Lobsters, confined to tight spaces in lobster traps for any length of time, commonly gang up and feast on weak lobsters and small fish (cod, pollock, flounders, sculpins, cunners, etc.) first tiring them out with repeated grabs before the final blows, and then devouring them from their extremities inward.

Microbes afflict lobsters with shell disease and the often-deadly blood disease "red tail" or gaffkemia, while herring and menhaden and, no doubt, some whales and sharks, filter-feed vast quantities of lobster larvae out of the water column. Seagulls also prowl the low tide mark of lobster nursery grounds for the tasty little ones, and cormorants pluck lobsters off of harbor bottoms. Further off, catfish or wolf fish consume lobsters, first crushing them with their powerful jaws equipped with front incisor and rear molar teeth. Striped bass, whose arrival in northern waters from the south also coincides with the lobster's peak molting and larval development stages, not only suck in whole lobsters at their once fairly safe shallows, but Brad Chase actually witnessed bass individually gulping down the lobster larvae at the surface. Chase once counted one hundred such larvae from just one bass's stomach sample. At fish processing plants during the commercial season for striped bass, fish cutters regularly fillet bass with bulging stomachs containing three to four lobsters. Vast schools of codfish gorge on lobsters living on the smooth bottom, too.

V

The Harvesting of Lobsters and the Big Changes

Today, lobsters are commercially harvested inshore and offshore, primarily by traps, gillnets, otter trawls and scallop dredges. The lobster trap accounts for about 90 percent of the annual United States and Canadian lobster landings. Small amounts of lobsters are caught by hand, too.

Non-Trap Harvesting Methods

Along the shoreline, scuba divers, snorkelers and the curious seasonally pluck lobsters off the ocean floor, especially during the summer and fall when they molt in close, and sometimes mass there right out in the open on sandy and muddy bottoms.

With the lobster being the common property resource it is, most states allow restricted recreational and commercial diving for lobsters, to the chagrin of most lobstermen. "Granted, they are everyone's lobsters in the ocean, but they are mine when they are in my traps," explains Robert "Mo" Morris Jr., a lobsterman from Rockport, Massachusetts. Many lobstermen, including myself, have had traps robbed by divers and have become suspicious of any diving activity near our pots. A lot of these divers use public boat ramps to launch their craft, and some even come by boat from out of state. Those few bad apples give the many good divers a bad name.

Lots of summertime youth energy is also devoted toward catching an occasional lobster by overturning rocks at low tide or diving down with a mask, snorkel and flippers.

Gillnets, otter trawls and scallop dredges also contribute to annual lobster catches, except in Maine and Canada where non-trap landings are prohibited. Other states, like Massachusetts, restrict non-trap lobster landings to one hundred lobsters per day per vessel up to a maximum of five hundred lobsters per trip, thus eliminating all other fishing boats from specifically targeting lobsters. Many lobstermen and regulators feel the mobile dredges and otter trawls also kill and injure a lot of lobsters.

Draggers like this one also scoop up lobsters from the bottom, either with nets or dredges.

Barrier-type bottom gillnets, made from transparent monofilament, are good at snagging lobsters on the move. Gillnets tied down as flounder nets are very efficient in catching lobsters. Back in the early 1990s, many Massachusetts flounder fishermen used to bait their tied-down gillnets with fish baits to attract lobsters during the spring and early summer. The baits were contained in bait bags, which were tied to the netting. No lobster limit existed then.

Lobsters also get scooped up by draggers that tow either otter trawls or scallop dredges. Some of these vessels are over 100 feet long and even powered by 1,500 horsepower diesels.

The rectangular-shaped steel scallop dredges, some fifteen feet wide, are good at digging into the bottom and picking up incidental catches of lobster and flatfish along with the scallops. The draggers and gillnetters often land the giant lobsters, which can weigh twenty-five pounds apiece or better.

On one of the offshore dragging trips I made on the *Judith Lee Rose* in the 1970s, three twenty-five-pound-apiece males and one thirty-one-pound male lobster, which had only one claw, were snagged. Before some states only allowed non-trap boats to land either one hundred lobsters per day or five hundred lobsters per trip, incidental lobster catches of two to three thousand pounds per trip or main lobster catches of ten to thirty thousand pounds weren't uncommon. Even the current five hundred lobsters per trip limit can weigh out to over two thousand pounds.

Remarkably, an occasional lobster gets hooked by longline gear, too. While longlining one spring off Cape Ann for cod, I caught a five-pound lobster that somehow got hooked

The Harvesting of Lobsters and the Big Changes

This Joe Sinagra drawing shows the setup of most of today's lobster boats, which are made from fiberglass, fall within the thirty- to fifty-foot-long range and are diesel powered. The traps are hauled aboard from a forward hauling station fitted with a davit and hydraulic pot hauler and are set back over the stern. Most of the boats are fitted with live lobster tanks, and some have pot wracks and rope wells on deck. Either one or two stern men or backup men often assist the owner with the lobstering.

in the knuckle of his big claw. Another time, my longline gear snagged a small lobster trying to swallow one of the hooked baits.

The Lobster Trap's Structure

The lobster trap catches the most lobsters landed. Although the trap's materials have greatly changed, its basic design and function haven't. In this day and age of environmental friendliness, the trap does not stir up (and possibly damage) the bottom like otter trawls and dredges do, but ironically, its lines and buoys sometimes entangle whales.

Anatomically, traps have a framework, nettings or "heads," weights and outer skin. The lobster trap is further broken down into "the kitchen" (usually the front end of the trap) and "the parlor" or "bedroom" (usually the rear of the trap). Sets of nailed on one-by-two inch bottom "runners" and one-by-one inch side "stiffeners" hold the

Pigeon Cove Harbor lobsterman Glenn Rose, assisted by stern man Dickie Crowell, hauls traps aboard his forty-foot *Marauder*.

typical wooden trap's frames together. Traps, depending on their size, can have three to four frames, or even half-round-shaped ones. The wire trap's framework is made up of a clipped-together box, often reinforced by corner braces, top bridges and two or three bottom runners.

For nettings, each trap has a set of side or kitchen heads, funnel-shaped nets usually with rings in their center holes, which allow lobsters entry into the baited kitchen area. Each pot also has a much larger and tapered parlor head, which not only guides lobsters into the trap's parlor section but also is designed to keep them from escaping here. Some of these lengthwise-running parlor heads also have rings in their openings. Those in the industry call these "hooped parlors". Some of these rings have lightweight plastic fingers that prevent lobsters from going back out through the parlor head rings. Many four-foot-long traps have two successive parlor sections, while other four-footers ("the double enders") have a parlor section at both trap ends with a kitchen section between them.

The trap's skins can vary from spaced-apart wooden laths to combinations of netting, vinyl-coated wire mesh and lath. The wire trap's skin is solely made out of wire mesh. The tops of the traps have a hinged door.

The lobster traps are weighed down by a balanced inner bottom arrangement of three to four bricks designed to make the trap land upright. Flat beach stones, window weights and a central slab of poured concrete work as trap weights, too. Some newly designed wire traps lack bricks as weights; instead, they have either bottoms made of heavy gauge wire mesh or runners constructed out of concrete which are either clipped, bolted or even molded onto the traps.

How a Trap Works

Today's trap works pretty much the same way it did one hundred years ago. Attracted by the bait, most lobsters will crawl into the trap's kitchen head forward and feed for a while. The big lobsters probably go in backwards with their huge claws crisscrossed the same way they return to their holes. The big lobsters are also more cautious about entering a trap than the little ones. Scuba observations of mine have shown me many lobsters will mill around the pot and crawl all over it before they go in, while others will go right in. The latter lobsters were probably conditioned to go in and out of traps thanks to the lobster gauge and escape vents.

While diving, I've witnessed a cautious big lobster come right out of its hole and go into a lobster trap once it observed a little lobster go into that trap and feed. The big lobster then exercised its dominance and immediately pushed the little lobster aside. Some lobsters will go right out the way they came in after feeding, while many will attempt to escape through the parlor head.

The lobster trap is far from being 100 percent efficient, but they do catch lobsters to the point of sometimes cleaning up all of the lobsters in an area, especially during the winter and spring, thus forcing a lobsterman to move his traps. In traps stacked ashore, accidentally captured blue jays, robins, cats, raccoons and even skunks will also try to get out through the parlor head.

Once in the parlor section, the lobsters will try to find a way out before settling in a corner—tail first—where they can better defend themselves. In most cases, lobsters are pretty much trapped in the parlor sections, unless they find just one break in the head's netting. Hauling an empty trap usually prompts the lobsterman to look closely at the parlor head to see if there are any breaks in the netting.

When feeding, lobsters will go right into the trap the moment it hits the bottom even in the broad daylight. I've re-hauled a trap just a few minutes after I've set it back only to find a lobster already in its kitchen area. When the water is cold, some lobsters take longer to leave their homes and go into a trap. Lobstermen must then use a "bait and wait" strategy for up to a week for best results. But it's not uncommon for lobsters in the shallow water during the winter to go into traps that have been set for just one night.

Many lobstermen feel a full trap, including one having crabs and small lobsters, will discourage other lobsters from entering. For this very reason, most lobstermen equip their traps' parlor sections with several small escape vents to assure most sub-legal lobsters and crabs easy escape. They also believe a bigger trap, which obviously has more space, will fish better than a small trap during long sets when there are more lobsters around.

Pigeon Cove Harbor lobsterman Fred Hillier, assisted by stern man Greg Chiancola, hauls traps during a damp spring day aboard his *Blueberry*, a thirty-six-foot-long Nova Scotian-style lobster boat.

How the Traps Are Worked

The inshore lobsterman's workday hasn't changed that much over the years, but the intensity and equipment have. We still go to sea in boats with fresh bait, sometimes helped by stern men, haul traps and then return to port with lobsters. Instead of going out for the day, offshore lobstermen and their four- to five-men crews generally "trip fish" up to a week at sea with more gear further off in larger vessels, many of which are company-owned. Lobstermen all have their routines and regular garden patches, where they know the topography and current patterns, storm threats there, and the yearly lobster movements. Here, lobstermen often work on memories of catches as to where the lobsters should be and when they will be there. Lobstermen also work under a constant irony best described by Al Koscinak of Gloucester, who runs Appledore III Charters out of Rockport. "The ocean is your friend one day, and the next day it wants to kill you."

The Lobsterman's Job

The lobsterman's job is to hunt down lobsters, trap them and bring them up from the bottom, and then enter these into the marketplace. The lobstermen are like the lions of the Serengeti Plains in East Africa; both predators seasonally follow their prey. In the case of the lions, their main prey is the wildebeests, or gnus. Since lobstermen most of

The Harvesting of Lobsters and the Big Changes

The *Flying Finn* sets back a string of freshly baited traps. The forward motion of the boat and the weight of the first trap in the water pull the rest of the traps off the boat.

the time can't see down to the bottom from the surface, they can only guess what goes on down there. Most of the time this guessing works as to where the lobsters are.

The Inshore Lobsterman's Daily Routine

The inshore lobsterman usually begins his day early in the morning, sometimes before daybreak. By then he has listened to weather reports, checked the weather out himself, had breakfast either at home or a local coffee shop, mustered up his crew if need be, put bait aboard and headed out of the harbor with a game plan. Insulated boots and rubber gloves, bib trousers, hooded rain jacket and hat keep him warm and dry during the cold months. In the summer, many lobstermen will use an apron instead of bib pants to keep them dry. Just about all lobstermen wear a hat year round, mainly to protect them from the sun and cut down on glare from the water. Most put their hats on as soon as they leave the house; this is second nature to them. Lobstermen are creatures of habit; we have our regular schedules. At Pigeon Cove Harbor, out of which I lobster, a group of lobstermen, including myself, begin at daybreak, while another group starts around 7:00 a.m., and there's the late group which begins around 9:00 a.m.

Once on the grounds, the fisherman will usually haul one-third to one-half of his total string, which can number up to around eight hundred traps in United States inshore waters (except for Maine's inshore Zone E, which imposed a six hundred-trap limit in the year 2000). Some men with smaller strings might haul all of them out each day.

Lobstermen frequently break their strings into eastern/western or northern/southern components. Aided by memory and at times electronic navigational equipment when out of sight of land, the fisherman knows exactly where and how his traps have been set, where the strings begin and end, and which ones he has hauled. Fog occasionally disorients a lobsterman, and it might trick him into hauling the same trap twice. Lobstermen also like to let their traps set two or three nights, and up to a week in the winter, before hauling them. The longer set strategy gives the traps more chance to fill with lobsters and makes the lobsterman's effort more cost effective.

Lobstermen fish their traps either as singles or as trawls. Individual preference, local tradition and sometimes area regulations dictate the choice. Single traps have only one buoy line, while a trawl, which is buoyed at both ends, has a series of traps tied into a ground line at regular intervals, often every sixty feet apart. Trawls not only produce fewer surface buoys and are quicker to haul than singles, but also result in less gear loss both by having a second buoy to get the string of traps back if the other buoy is lost and by clinging better to the bottom during storms. But some lobstermen consider fishing single traps to be more efficient than trawls since each single trap can be placed on selected bottom.

Cape Ann lobstermen commonly fish five, ten and twenty-pot trawls, while Long Island Sound lobstermen work six- to eight-pot trawls. Offshore lobstermen's trawls commonly each contain twenty-five to forty-four pots with approximately 180 feet between each trap. Outer Nova Scotia's Area 33 lobstermen fish single traps and many of the coastal Maine lobstermen fish two-pot "bull" trawls.

To prevent crisscrossing of gear, most trawl lobstermen set their trawls in agreed-upon patterns ranging from parallel to the shore, to north to south or east to west. In addition, a special buoy scheme designates which end of the trawl is which. Single buoys, some having a flag on the tip of their sticks, mark the north and western ends, while the opposite end buoys are usually doubled as either two on a single stick or two separate buoys tied together, including a high-flier buoy with a flag and radar reflector on its stick and a tide ball.

Once at the beginning of the string, the lobsterman will maneuver his boat to the buoy and usually gaff it aboard, and then run its rope over the davit's lobster block and through the pot hauler before yanking up the trap, sometimes so fast the rope will "sing." Lobstermen regulate the rate the hauler pulls by opening and closing a hydraulic valve. Every lobsterman has his own buoy color, ranging from all one color to combinations of two and three colors. The owner's lobster license number and first initial and last name or his initials are branded into each buoy. Traps have colored tags bearing this same information. Area lobstermen know which buoy colors belong to whom.

With the trap aboard, the lobsterman and stern man or backup man will first empty the trap, keeping the potentially-legal size lobsters and tossing back any "shorts" (little lobsters), "seeders" or V-notch females. He will then rebait it before setting the trap back, usually in a slightly different location, and advancing to the next buoy. Most lobstermen are careful

The Harvesting of Lobsters and the Big Changes

Many lobstermen, like Bob Morris III of Rockport, Massachusetts, check their traps in skiffs. Here, Morris heads to sea in his eighteen-foot-long skiff.

not to set their trap back on anyone else's. Lobstermen have a very good eye of what is legal size. Sometimes a little lobster that looks undersized will fool you and be legal. The opposite is sometimes true for a bigger-looking lobster that has a short carapace.

The legal lobsters' claws are first banded with rubber bands before they are stored and kept alive in either on-deck or below-deck wet lobster tanks containing circulating seawater. Lobstermen use a special banding tool, resembling a pair of pliers, to band the lobsters' claws. Bands have pretty much replaced the traditional wooden pegs, which would puncture and thus weaken the lobster. Questionable legal-size lobsters are measured by a gauge. Most fishing areas have a $3\frac{1}{4}$-inch minimal size carapace length, while Area 1 (off Maine, New Hampshire and Massachusetts) has a maximum 5-inch size. Lobstermen also check for egg bearing and V-notch females, which are illegal to take. Most lobstermen feel "these are our future" (like the shorts). The V-notch females have a V-notch on the second flipper in on the right side of their tails. They were notched by either fishermen or regulators while previously carrying eggs. Lobster laws now require many lobstermen to V-notch all egg bearing females before returning them to the ocean.

At day's end, the lobsterman will usually sell his catch to a dealer before returning to the mooring. Some lobstermen belong to cooperatives and are required to sell their entire catches there, while others are suppliers at will and can sell to whomever they want. Other lobstermen are required to sell to a specific dealer because of docking

Gloucester, Massachusetts, lobsterman Nick Parisi, accompanied by his dog, hauls traps in his eighteen-foot skiff.

arrangements. The dock-to-dock time required for many lobstermen to go out and haul three hundred traps frequently takes about six to seven hours. Some days go according to schedule, while others can take longer, especially if the weather is bad.

Lobstermen will usually haul half of their string one day and the other half the next time out, and so on. The lobsterman's season is made up of such repeated daily ventures. The ventures add up by year's end; you get out of the business what you put into it. Lobstermen, like all fishermen, are paid by the pound and not the hour. Daily inshore catches can range from a couple of buckets (approximately 40 pounds) to 1,200 to 1,500 pounds.

The opening days of Canada's approximately twenty fishing districts often produce huge landings; some boats land two thousand- to five thousand-pound catches for the day. On trap "dump day" (opening day) frenzied fishermen head out to sea with their boats loaded with traps, often over three hundred of them, set the traps, and then turn right around and haul these once or twice. Bad weather has occasionally forced government officials to delay the opening day. Boats have been lost on trap days due to bad weather. One young Canadian fisherman landed 160,000 pounds during one of his recent open seasons around 2001. Many inshore lobstermen earn gross incomes in the six-figure range. Stern men in the big lobster-producing zones of Maine can earn $65,000 for seasonal work. Every cove and harbor has its high-liner or high-liners. "Lobstermen cry a lot, but we keep getting new four-wheel drive trucks and new boats," says one fisherman.

Offshore vessels with four- to five-man crews typically make four- to seven-day-long trips. Sometimes they have to travel twenty hours or more to reach their grounds, which

The Harvesting of Lobsters and the Big Changes

Many of today's lobster boats have gotten bigger and faster. Here the forty-five-foot-long by eighteen-foot-wide Nova Scotian-style lobster boat, *Jane & Girls*, owned and operated by Trevor Malone of East Pubnico, Nova Scotia, cruises along at twenty-eight knots. The boat is powered by an eight hundred horsepower Cat diesel.

include Georges Bank and its outer canyons. The offshore lobstermen are especially dependant upon charts and electronic navigational equipment to find their gear. They record where each trawl has been set. During each trip, the crews typically haul about five hundred traps daily, with the goal of going through their whole string, which can number two thousand traps, at least once, and sometimes parts of it twice if the catches have been good. Most of these boats have refrigerated seawater tanks below decks to keep their catches alive throughout the year.

The offshore lobsterman's best fishing often coincides with that of the inshore lobsterman's. Some four- to seven-day-long offshore trips have produced catches from twenty thousand to thirty thousand pounds. The offshore fishery got underway in the late 1960s, and trap boats and draggers tapped into this virgin resource. Since then, states have either banned draggers from landing lobsters or have limited their catches to one hundred lobsters per day or five hundred lobsters per trip.

Lobster Baits

Nowadays lobster baits can vary from fresh, salted and frozen, whole fish and skeletal forms (racks), to unconventional baits like animal hides. Just about anything will attract lobsters, except ocean catfish or wolf fish. Lobsters are instinctively petrified of this

Pigeon Cove Harbor lobsterman Mike Wayrynen, and to his left stern man Ollie Lahtinen, empty and re-bait a trap aboard the *Flying Finn*. Notice how they are dressed for summertime work.

predator, which crushes lobster and other shellfish with their powerful jaws and impressive incisor and molar teeth.

Many years ago, an old-time Cape Ann lobsterman even baited his traps with crushed-up short lobsters. Smashed crabs work well, too. I once even caught lobsters in a trap baited with an old pork chop. There have been stories of lobstermen using oil-soaked rags and bricks for bait, too. I can't understand why one would do this since the oil would go right into the lobster, giving it both an oily smell and taste. I've seen this happen to lobsters accidentally immersed in oily bilge water.

Bait preference is area-specific, although most lobstermen generally believe oily fishes like herring, mackerel, bluefish and menhaden work the best. Unfortunately, seals also are attracted to these oily baits and end up stealing many of these from the traps. Lots of inshore lobstermen from Cape Cod north to the Gulf of Maine prefer herring; those in Rhode Island bait with skates; the Long Island Sound fishermen use whole menhaden or "bunkers"; and the Canadian's main baits are whole mackerel, redfish racks and some herring. It's estimated that U.S. lobstermen annually use sixty thousand or more metric tons of herring for bait.

Many lobstermen along the Maine coast net alewives in rivers and streams in the spring for bait as these anadromous (going from salt water to fresh water and vice versa) fishes, also called freshwater herring, depart the ocean to spawn in freshwater lakes and ponds. They return to the salt water in the fall.

The Harvesting of Lobsters and the Big Changes

Essex, Massachusetts, lobsterman Ron Hemeon, and to his left stern man Sergio Garcia, represent some of the faces in the lobster industry today. Here, they stand alongside part of their day's catch aboard Hemeon's forty-two-foot-long *Alexis Margaret*.

Offshore lobstermen use the racks of redfish and groundfish like cod, haddock and flounder, since these hold up well. Many lobstermen also use trap-caught crabs and fish, especially cunners and sea ravens, to supplement their baits.

With groundfish and redfish, lobstermen generally spike these onto a hook. Others use a needle to string the bait onto a line. The softer herring, mackerel and pogy baits are generally bagged into special bait bags made from small-mesh netting. The warmer the water and swifter the currents, the quicker the bait disappears, sometimes only lasting two days. There is also a correlation between lobster catches and the rate at which the bait disappears in the traps. Usually, the faster the bait disappears, the better the lobster catches.

Animal hides, usually cut into four-by-four-inch pieces, have been slowly gaining acceptance. After curing several days, one piece will usually fish for about three weeks before needing replacement. This hide bait saves lobstermen rebaiting time, while it also keeps the traps fishing during bad weather when other baits will normally wash away. Lobstermen like to use hide baits once they begin fishing off the hard bottom edge, usually around September. The hide baits also repel most sand fleas, crabs and seals. In the fall on sandy bottoms, crabs often fill traps baited with regular baits and discourage lobsters from entering. Some lobstermen like the hide bait while others don't. A few users tell other lobstermen the hides are "no good" because they work so well and they don't want the others to catch on.

Many in the lobster industry believe the use of hide baits could create a negative public perception for lobsters. In this business, public perception is critical. Others also worry about adding something that didn't originally come out of the ocean to the ocean.

SeaLure and Smart Bait are two prominent companies that provide the bait hide. Many entrepreneurs with connections to slaughterhouses also sell hide baits by the five-gallon bucket out of their pickup trucks. They charge about thirty-five dollars per bucket. Each bucket contains about one hundred baits.

Bait is big business. Lobstermen say, "You gotta have it." Bait costs can add up to about 15 percent of the lobsterman's annual gross earnings. Cape Ann lobstermen pay around eighteen to twenty-two dollars per tote of herring (about one hundred pounds); one tote of herring generally baits around one hundred traps. Many boats spend approximately forty to one hundred dollars per day just on bait.

Numerous companies along the east coast specialize in fresh, frozen and salted baits. It's estimated that out of the annual 110 metric tons of herring caught in the United States, up to 70 metric tons are used as bait. Lots of lobster dealers, who offer one-stop service to their fishermen, also provide bait. Other lobstermen, like myself, get their bait independently either by truck or boat from a processor, usually the morning we are going out fishing. Many of us have bait coolers which allow us to stock up on available bait and have it on hand at all times.

Purse seiners and mid-water trawlers harvest much of the herring and mackerel lobster bait. Purse seiners also catch the menhaden. Draggers provide processors with most of the groundfish and redfish whose skeletal remains later become available for bait. Gloucester, the big fishing and processing port it is, has blessed Cape Ann lobstermen with readily-available bait sources, including mackerel and herring and fish gurry (bluefish, salmon, cod, haddock, flounder and redfish racks).

Bait shortages have occurred during past summers when the combined demand of lobster, tuna and striped bass fishermen has exceeded the supply, often made short by either the warm water driving away or making the cold-water-loving herring hard to catch. These bait shortages have resulted in payless days for fishermen. Hide baits have eased the dependence on fish baits for many lobstermen.

Back in the 1950s and '60s, bait was cheap and plentiful, especially in Gloucester. I can recall having the pick of redfish plants there then. You stood alongside of a conveyor belt with your bucket or barrel and simply plucked redfish off of it. The bait only cost one dollar per bushel.

Lobstermen also used to catch some of their lobster baits then. I once bait fished for cunners with special cunner traps and hand-lined for mud hake or squirrel hake summer evenings. Shoreline fish traps used to seasonally provide fresh baits—small pollock, mud hake and herring—too. Other lobstermen netted mackerel, herring and pogies with surface gillnets.

The Harvesting of Lobsters and the Big Changes

Pigeon Cove Harbor senior lobsterman Dickie Olson gives the thumbs up as he enters his fish shack.

What A Trap Catches

First and foremost a trap catches lobsters. Tourists often ask, "How many lobsters can a trap catch?" Trap catches, commonly measured in "pounds per pot" (the fishermen's version—total catch weight divided by the number of traps hauled), can vary from day to day, area to area, and season to season. As stated earlier, approximately twenty-five pounds of lobsters (seven big ones in this case) was my largest single trap catch, caught by a thirty-six-inch-long wire trap one spring. Bro Coté reports his largest offshore catch for a twenty-five-pot trawl of four-foot-long traps was "2200 pounds." One to two pounds per pot is considered good fishing; catches can range from only a third of a pound per pot way up to four to five pounds per pot. Not all traps come up with lobsters in them. Traps that come up empty of lobsters always seem to be heavier than those with lobsters. Today's traps are fitted with escape vents, and they release most of the shorts or crickets. Years ago, the little ones commonly jammed some traps' parlor sections, especially during the Fourth of July shedder runs. This wasn't good for the stock since many of these critters crush one another in the traps.

A seventeen-pound female was the largest lobster I've ever caught in a trap—this time a Casco Bay style wire trap with seven-inch diameter side head rings or hoops. The big lobster somehow managed to squeeze into the trap's kitchen section, probably crawling in backwards (through the ring) with its claws extended and crisscrossed.

In the 1970s I also caught a nineteen-pound male lobster while I was scuba diving one August day in Ipswich Bay. That day I swam about ten feet off the bottom, and there

LOBSTERING OFF CAPE ANN

Gloucester, Massachusetts, lobsterman Arthur Surrette returns to port aboard his *Tempest Tossed II* after a day out fishing in the winter.

The Harvesting of Lobsters and the Big Changes

One of the most unusual creatures ever, a giant water beetle, came up in one of Dave Foote's lobster traps set in Gloucester Harbor.

was this perfect male lobster sitting right in the open along the edge of rocky and muddy bottom forty feet down. I grabbed the lobster by the backs of its claws and surfaced with it. Nova Scotian lobsterman Jerry Himmelman recalls a lobsterman friend of his who caught a twenty-three-pound lobster that came up on the topside of a trap. The creature's walking legs got stuck in the spaces between the laths.

What Else Do Traps Catch?

Beside lobsters, a trap can catch a variety of other invertebrates and vertebrates: fish, birds, and even mammals. This is my career's non-lobster catch. The invertebrates range from sponges to sea urchins and starfish to crabs, to sea scallops and common periwinkles. A foot-long sea lamprey attached to the sides of a codfish even came up in one of my traps. Other trapped fishes over the years include spiny dogfish sharks, flounders, mackerel, cusk, catfish, tautog, lumpfish, sea snails, pipefish, trigger fish, scup and black sea bass. Sad to say, occasional double-crested cormorants pursuing small fish nibbling on the bait also get trapped and drowned. Hard to believe, raccoons and feral cats go into lobster traps at dead low tide after the bait and also get trapped and drowned. Rats also go into exposed traps, but quickly chew the netting to free themselves.

The most unusual lobster trap catch that I've ever heard of and seen was a giant water bug. During October of 2003, Gloucester lobstermen Dave Foote and Bobby Oliver had a four-inch-long giant water bug drop out of one of their traps set in

Monhegan Island lobsterman Mathew Thomson chats with a friend during the Maine Fishermen's Forum.

Gloucester Harbor thirty feet down. Dr. Lanna Cheng, a noted marine entomologist at Scripps Institution of Oceanography in San Diego, identified the creature. "They live among vegetation in freshwater ponds and streams. I suspect your specimen was probably attracted to light at sea and became stranded there. They don't normally fly except when they are ready to disperse," she says. This specimen probably came from a nearby pond and then was attracted to the bright lights along the shore before being blown out into Gloucester Harbor and sinking. Cheng added, "Most aquatic insects can deal with salt water. Salt water won't kill them. Their very thick cuticles protect them. This animal would have to come to the surface to breathe."

How Traps Are Lost

Man's Actions

Traps are lost by both nature and man. Man's actions can be either deliberate or unintentional. Snarls, especially during the warm summer months, can incite some short-tempered lobstermen to hack away others' buoys, lines and traps in order to free their own gear. Most areas have a handful of hackers who are well known to the other lobstermen. Other hackers, through maliciousness, jealousy, insecurity, or even self-preservation will

The Harvesting of Lobsters and the Big Changes

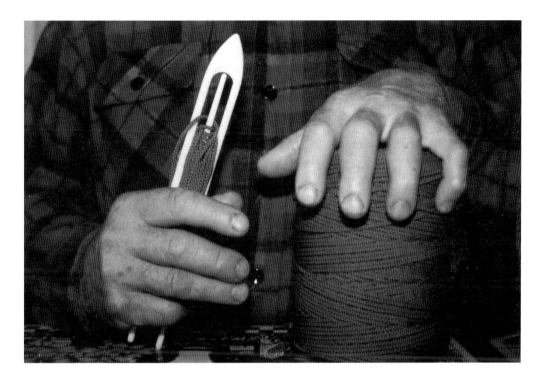

Lobstermen's hands and some of the tools of the trade: a knitting needle and a ball of heading twine.

destroy competitors' gear, especially newcomers'. They do this to either drive them out of a fishing area or out of business completely. When I began lobstering in 1960, you had to fish within your town's or city's waters or risk having your gear cut away. This isn't so common any more off Cape Ann since many of its lobstermen's ranges now encompass several townships. Numerous Maine fishermen are still fiercely territorial. Lobstermen have also had their traps stolen at sea and on land while in storage, usually by other lobstermen who desperately need traps.

Gear cutting can spark "lobster wars." An Ipswich Bay lobster war occurred in the late 1960s between an independent old-timer who has since passed away and several young fellows. The whippersnappers believed the old-timer was hauling and stealing their lobster gear. Well, someone cut a few of the old man's traps, and he did the same to the young lobstermen's gear. The feud escalated to boat burnings and sinkings before the authorities stepped in and prevented possible loss of life. Tempers then ran high enough for murder. Fortunately, most lobstermen realize "there are no winners in lobster wars," and these rarely happen. "Fishermen [including fixed-gear versus mobile-gear fishermen] should never fight amongst themselves; the real enemy is the government," says one fisherman.

Lobster gear sometimes gets lost when it is towed away by mobile-gear fishing vessels like scallopers, bottom draggers and mid-water trawlers. Many lobstermen today fish on a lot of traditional dragger bottom. Most of the mobile vessels will try to avoid fixed-

Many lobster boats today, like the *Lady Elaine* based out of Pigeon Cove Harbor, have the power to haul up bunches of twenty-five to fifty traps with their hydraulic pot haulers. Here *(left to right)* Justin Knowlton, Toby O'Connell and his father Captain Dennis O'Connell work on a bunch of traps.

gear conflicts. Getting snarled in lobster gear affects their gear's fishing effectiveness, and it can damage their nets, too. Many unintentional conflicts occur during the dark when the buoys cannot be seen and also when strong tides keep the lobster buoys under water. Although illegal, a few mobile-gear operators will deliberately tow their gear through lobster traps if they are in their way. "I had $43,000 worth of gear towed away during one night," recalls former offshore lobsterman Stevie Robbins II, of Little Deer Isle, Maine. This shows how expensive losses in gear can be.

Recreational and commercial boat traffic also chop away a lot of lobster buoys, and in some cases even tow traps away. In 1998, Pigeon Cove Harbor lobsterman Bob Morris Jr. recovered a two-pot trawl belonging to York Harbor, Maine, lobsterman Jeff Donnell in the shipping lane several miles off the Rockport coast. Donnell's buoy probably got stuck in either a tug's rudder or on its barge before being towed a good thirty miles from York, Maine, to Rockport, Massachusetts, where the buoy line finally parted and the traps got snarled with Morris's. Morris later returned the gear.

Recreational fishermen occasionally cut lobster gear to free their anchors, hooks and plugs. Scuba divers even sometimes cut the buoy line off a trap or two inshore during the summer and visit and rebait those traps weekly to get lobsters.

The Harvesting of Lobsters and the Big Changes

Small draggers equipped with hydraulic net reels, like the *Special K* shown here, can haul up bunches of one hundred traps or more. The *Special K* has raised a big bunch of traps and is towing it shoreward in this shot.

Lobstermen unintentionally lose their own traps either through haste or carelessness at sea and even on land. Some lobstermen, including myself, have been in such a hurry to set traps at season's start that we've pushed them overboard without tying lines to them. You realize this the moment you push a trap over, but by then it's too late. An occasional lobstermen has also cut his lines to free a snarl and then forgotten to tie them back. Others have not tied down loads of traps on aboard and have lost these when their boat suddenly rolled or have had to suddenly jettison a large load of traps aboard his weighed-down lobster boat that is now taking on water or risk sinking. Big storms have even washed away or damaged traps stored on wharfs. I once cut down a big tree in my back yard, and this fell the wrong way—right on some brand-new traps, which were pancaked beyond repair.

BOTTOM SNAGS

Bottom snags have been another source of trap losses. The big crevices and caves of the Sandy Bay breakwater seem to gulp down traps even in calm weather. Diving is the only way to get back many of these pots. I've often wondered how some of the traps ever get into some of the narrow opening caves and crevices in the first place. Freeing them often

Lobstering off Cape Ann

Northeasters also often wash ashore huge globs of snarled rope and smashed traps.

involves reaching down and pivoting the trap from one corner and then rocking it back and forth and gradually lifting it out. While diving, I have not always been able to free all of the traps there. On one occasion, while attempting to free a trap along the inner breakwater edge near Avery'd Ledge, the force of a breaking roller catapulted me a good one hundred feet.

Traps get snagged on shipwrecks, anchors and unmarked ghost lobster gear, too. The rusted, jagged remnants of the old Liberty ship's steel hull behind the Sandy Bay breakwater not only hangs down pots but also cuts their lines when tugged on.

During the 1970s I helped salvage a 2,500-pound anchor 40 feet down behind the Seaside Cemetery in Lanesville that was draped with lost traps. This anchor now sits in front of the Ralph B. O'Malley Middle School in Gloucester.

Nature's Actions

Storms destroy a lot of lobster gear. For most areas, the northeaster is the biggest destroyer of traps. Lobstermen constantly keep their senses tuned in to the weather. Just the mere mention of a northeaster coming will silence a lobsterman in any setting. He will immediately wonder how bad the predicted storm might be and whether his traps will be safe.

Dave Eastman and Ron Parnell, two veteran Ipswich Bay lobstermen, have been burned, like most of their peers, by northeasters of different intensities over the years.

The Harvesting of Lobsters and the Big Changes

Redfish is one of the preferred lobster baits today. This bait not only lasts a long time, but its natural oils attract the lobsters.

"The news of an impending northeaster puts you in a foul mood, and you get that sick feeling inside," Eastman says. "You also do not know whether or not to believe the weathermen at first. You always hope they will be wrong, and the storm either won't materialize or it will go out to sea." Parnell adds, "You just feel ugh, and you say to yourself, 'Here we go again.' But sometimes the weathermen are wrong. I remember one storm that was supposed to be a fast mover, and it lasted about a week."

Once the storm seems a sure thing, lobstermen begin relocating their traps, especially from the surf-prone shallows, to storm-safe areas in deeper water on smooth bottom off the edges, usually two to three days before the northeaster's anticipated arrival. Although the traps have a better chance off the edges, big northeasters sometimes sweep them away from here. The completion of this backbreaking gear shifting task usually brings about a sense of relief.

With the high cost of fishing gear today, most lobstermen prefer to play it safe rather than be sorry, knowing a false alarm only costs them some fishing time, fuel and sometimes sore muscles. Frequently, all this gear moving is for nothing.

While dealing with the actual storm, lobstermen often observe it and speculate on how bad it might be. Doing so helps calm their nerves. Many of the men repeatedly drive about in their trucks to their favorite observation spots at beach parking lots, their coves and along coastal roads to look for special clues that tell them a storm's intensity. These include dead fish washing up on beaches, wave heights and steepness, high tide durations, amount of storm surge, signs of trap movement, color of the water and whether or not cobbles roll in the crashing surf at certain beaches. Most lobstermen can look at a stormy ocean and their gut feelings are, "That's rough" or "That's not bad at all."

"While you sit there in your truck alongside the ocean you feel helpless, and you think of how badly the traps might be getting damaged," Eastman says. "You also hope for the best. That's all you can do."

"You don't look forward to going back out there," Parnell adds.

Although most northeasters won't destroy a man's entire string of traps, they often create a lot of necessary work to get the string back fishing, including trap unsnarling and repairing, freeing those that are hung down, replacing lost ones and emptying some of mud and seaweed.

Northeasters frequently make some lobstermen further enjoy the securities of their homes and family. Many even monitor aspects of the storm, especially the top wind speed, with their own weather instruments. Parnell says he "turns up the TV volume to block out the sound of the wind."

One feeling never seems to leave lobstermen. "You know a northeaster is going to happen again," Eastman says. "The question is, how long in between?"

The wave actions of the northeaster often go down deep, sometimes so strong that the stirring action literally brings the bottom up to the surface as seen by brownish-black water. These storms can also destroy marine and human life, flood and damage the coast, and even alter the intertidal zone. These commonly moisture-laden, unpredictable storm systems have plenty of water space off the east coast to crank up sometimes hurricane-force onshore winds. The big northeasters pile up water inshore that turns low tides into high tides, creates storm surges and riles the ocean's top and bottom. The storms use these ingredients and the regular ebbing and flowing tides to weaponize the weight and motion of water, which both pummels exposed shoreline with bowling ball impact and scours the bottom with raging rivers. These rivers, especially in the strong tide areas, sweep away the traps—both singles and trawls—bouncing many along the bottom and carrying others off the bottom. The pots keep moving until either the wind abates or they get snagged, often against a rocky outcropping or another hung-down trap, and then the storm plays ring around the rosy, winding them ever so tightly into a Christmas tree-shaped mound and their buoy lines into a tree trunk. On muddy bottoms, these bunches sometimes get carpeted by fine black mud.

Fortunately, lobstermen have the skill (and the thirty- to fifty-six-feet-long diesel-propelled boats with powerful hydraulics) to deal with most bunches. The lobstermen then get their turn to weaponize hydraulic fluids to fight the storm's aftermath. Boats haul up the snarls with their pot haulers, either alone or in teams, while lobster boat and dragger combinations use their stern A-frames and net reels. The pot hauler, which can only haul in one line at a time, "will keep hauling until the rope breaks," explains Dennis

The Harvesting of Lobsters and the Big Changes

Sea urchins frequently come up in lobster traps. Their spines can cause long-lasting, painful wounds.

O'Connell, owner-operator of the approximately forty-foot lobster boat/dragger *Lady Elaine*. This line is often under so much strain in the pot hauler that it will sing.

The net reels, which can pull in many lines at once, minus their buoys, are generally easier, safer, and better able to lift heavier loads—over four thousand pounds—than the pot haulers. "My biggest bunch so far had seventy to eighty traps," says dragger-man Dustin Ketchopulos. Logs and ghost gear lost years ago frequently come up in the bunches, too. Ketchopulos (with his fifty-six-foot *Special K*), O'Connell and lobstermen/dragger-men Jay and Peter Van Der Pool (via their forty-four-foot *Allysha & Andrew*) have hauled up and brought in hundreds of snarled lobster traps for their grateful fellow lobstermen. Other lobstermen often help these men during the operation by ferrying in hauled-up traps.

Hauling up the bunches frequently pushes the equipment and the fishermen's temperaments to their limits. A bunch can take a non-alert fisherman right to the bottom in a split second if a line suddenly snaps and his feet either get entangled in deck line or his hands get snagged in some jagged trap wire. A four-foot trap weighs about fifty pounds, and these are often filled with mud, too. The December 6 and 7 "Classic Storm '03" created the biggest bunch I've ever seen, containing approximately one thousand traps, one hundred feet down off Rockport. This snarl has since been hauled up a little bit at a time.

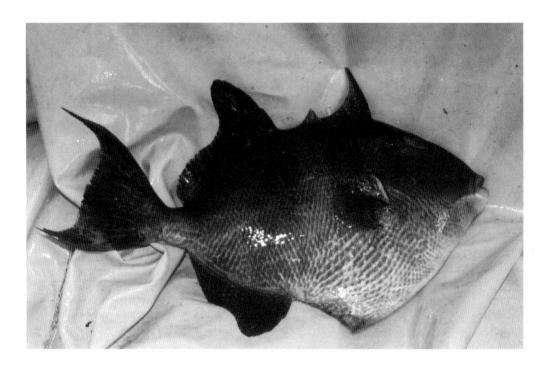

Storms sometimes either wash in or up the coast southern species like this triggerfish, which came up in one of my traps. I placed this specimen in a bucket of water. While playing dead there, it nearly bit the tip of my finger off as I reached in to stir it.

Years ago, some scuba diving friends and I used to retrieve some of these bunches. We would go down with a line, also attached to a surface lobster boat, and cut off several traps at a time and tie these to the line. The crew would quickly haul up the line, yank aboard the traps, and then lower back the line. In the spring lumpfish often lay long strings of pinkish eggs along some of the bunches, and their gray or reddish-colored parents stand guard.

Back in the 1950s and '60s, the late Annisquam Harbor lobsterman Winthrop "Bunt" Davis hauled up many bunches of wooden traps with his thirty-six-foot-long dragger/lobster boat *Carolyn Rene*, using a mast and boom, blocks and tackles, and the big winch head of his powerful drum winch. A gang of other lobstermen would help Davis, either aboard his boat or alongside in their boats. Davis would first tow the bunch free and then begin the hauling process by wrapping a strap around the rope trunk and then gradually hoisting, tying off, hoisting and so on. Once the traps broke the surface, they would be cut off, brought aboard and transferred to one of the waiting boats. All of the gear, much of it undamaged, would be brought into one of the nearby coves to be retrieved by their owners, who also untangled their ropes and buoys from the snarl.

One such day's activities remains clear in my mind, and it still makes me chuckle. Davis and another lobster boat, the *Chicawa*, were towing a huge bunch of one hundred

The Harvesting of Lobsters and the Big Changes

traps from Halibut Point to the sheltered shallows of Folly Cove beach. Here, they would haul up the snarl. Unbeknownst to everyone, the Feds had this same cove under surveillance for marijuana smuggling. While towing, Davis had called other lobstermen over the radio. "We are working on pots, and we need your help." Within an hour, three U.S. Coast Guard cutters with their lights flashing and with armed Guardsmen standing along their rails converged on the boats. Davis was not happy at all when Guardsmen came alongside while he was dangerously hauling up a bunch and asked him to "stop what you are doing and prepare for boarding." An angry Davis told the men, "We are working on *pots*—*pots*—lobster pots, not pot." The surprised Coast Guardsmen soon realized their mistake. Years later, a party who had rented a house overlooking Folly Cove was arrested for drug trafficking.

Strong offshore tides created by either Gulf Stream eddies or full moons sometimes pull lobster buoys right down to the bottom, especially offshore. The intense bottom pressure will even shrivel many of the buoys. Fortunately, with today's electronics, which can pinpoint positions, lobstermen keep coordinates of where their gear was last set, and they can grapple most of their gear back, although this often takes a long time.

A storm doesn't even have to pass nearby to wreak havoc on inshore lobster gear. Seas from offshore storms travel hundreds of miles, eventually pounding the shoreline as heavy surf. I still remember hearing the thundering noise coming from the shore one calm and sunny May afternoon in 1965. "Where did that come from? We've had no storm," I said. That offshore storm's seas washed ashore about fifty of my traps.

Most lobstermen prefer to be safe rather than sorry; they react to storm threats by moving their gear out to deeper and safer water. Traps usually set in 150 to 180 feet of water or greater are pretty safe. But Bro Coté remembers that the October 31, 1991, No Name storm moved one of his 40-pot trawls set along the 300-foot-deep edge of Georges Bank "ten miles to the west." These traps were neither damaged nor snarled.

The increased use of wire traps (which are heavier in the water, offer less resistance to moving water and can take more of a beating than the old-style wooden traps) has dramatically cut down on trap storm damage and losses. But a bad storm will even destroy these durable traps. The October 1991 and December 1992 mega-storms cost me 667 wire traps which either got sucked out, never to be seen again, or totally destroyed. Most of these 4-foot-long, 2-inch mesh wire pots, even weighed down with 18 pounds of cement bricks, washed ashore in huge mounds of rolled up basketball-sized traps, snarled rope, broken buoys and seaweed. The 1991 storm moved many of these traps, set in over 60 feet of water along the edge of rock and mud, three to four miles before washing them ashore, sometimes in bunches of 25 traps. The December 6–7, 2003, storm even moved traps set in 46 fathoms (approximately 260 feet). "That's the first time I've ever had that happen," says Dennis O'Connell.

Trap losses and their lost fishing time are always painful and expensive. There have been cases where I've never gotten to haul traps that were just set because a storm struck

James Knott Sr., president and CEO of Riverdale Mills Corp. at Northbridge, Massachusetts, stands alongside rolls of his Aquamesh trap wire which wait to be vinyl coated. Knott was the first man to put a wire lobster trap in the Atlantic Ocean. His products now go all over the world.

and destroyed those traps. Today a typical completed thirty-six-inch-long wire trap with a rope and buoy costs up to $70, while a four-foot-long one with rope and buoy can go for $90 to $100. A ten-pot trawl of four-foot-long wire traps complete with rope and buoys is worth about $750 to $1,000. A Gloucester lobsterman getting started in 2004 spent $70,000 for eight hundred new wire traps, along with their ropes and buoys. Not including their boats and trucks, lobstermen are truly sitting on a big $30,000 to $60,000 gear investment.

Most storms will destroy only a portion of your string; the real damage results in snarling and prematurely aging a lot of the gear by chafing its ropes and wearing off the vinyl coatings of the wire traps. Many lobstermen, including myself, have spare traps at home ready to go at all times. Today's lobster trap factories can manufacture new traps quickly and will deliver them right to the dock. Backlogs can occur when everyone wants new traps and gear at once.

Ironically, lobstermen occasionally get back lost gear from some storms that either wash ashore such traps or snarl them into bunches that later get hauled up. I particularly remember one fall northeaster that actually washed ashore three traps that I had lost the previous summer. The traps were in good shape, too.

Another big irony associated with storms is no matter how intense a storm has been, the sun will eventually come out, tranquility will follow and life will go on. This can be a hard pill to swallow after a storm has either destroyed your boat or, more commonly, a good portion of your traps, leaving your business in limbo. The big December 25, 2004, storm "put an end to my lobstering world and left me with only uncertainty," explains Pigeon

The Harvesting of Lobsters and the Big Changes

Cove Harbor lobsterman Mike Mattson, who lost his thirty-two-foot lobster boat *Lower Argyll* in that storm. The boat sank at its mooring and later smashed up on the bottom.

Besides the moisture-laden low-pressure system northeasters, there are also dry northeasters that move down from the north as backdoor cold fronts. These dry northeasters can be ferocious windstorms, but usually after five or six hours the winds quickly abate. As their name suggests, these systems pack no moisture, and the sun also shines during their presence.

Although a wire trap can last a lifetime if fished just seasonally on sheltered smooth bottom, most last four to six years. Gloucester lobsterman Tim Sullivan is still fishing wire lobster traps in 2005 that he made in 1988. Some lobstermen, wanting to have the best gear possible, automatically sell their three- to four-year-old traps. Others will keep repairing their pots. Selling used traps has become a business for some trap dealers. Many new trap manufacturers also take used traps as trade-ins and later sell them as well as with new ones.

The Big Changes

Since the 1960s, the wire trap, hydraulic pot hauler and fiberglass lobster boat have helped revolutionize the trap fishery, leading way to improved lobster gear, greater fishing time and more efficient, profitable and easier operations. The revolutions were initially met with great resistance.

THE LOBSTER TRAP REVOLUTION

The traditional wooden lobster trap has been largely pushed aside by the easier-to-handle, more durable and efficient wire trap. Changes in the wire trap are still occurring, especially with its accessories.

Bulletin of the United States Fish Commission Vol. XIX, historical records of the Maine lobster industry, describe how the lath trap came about during the late 1800s as a replacement to the baited hoop nets.

> *As there was no way of closing the mouth of the pot after a lobster had entered, these nets had to be constantly watched… The fishermen would generally go out in the evening and at short intervals he would haul in his nets and remove whatever lobsters they might contain. The constant attention necessary in attending to these hoop nets led the fishermen to devise an apparatus which would hold the lobsters after once entering and would require only occasional visits and "lath" traps were found to fulfill all requirements. They are usually about four feet in length…and in Maine are usually of semi-cylindrical form.*

When my lobstering career began in 1960 at Lane's Cove on Cape Ann, around fifteen lobstermen seasonally worked their fifty- to one hundred-trap strings along the shore in either hand-rowed dories or in powerboats with winch heads. They fished half-round

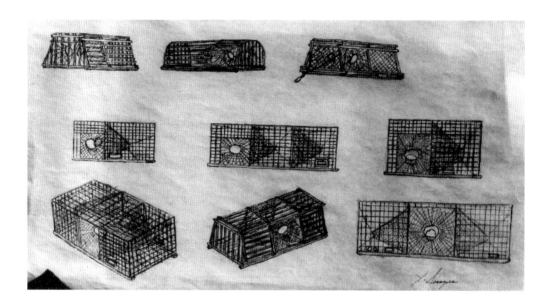

This Joe Sinagra drawing illustrates how wooden trap styles have evolved into today's different types of wire traps.

and rectangular-shaped wooden traps around thirty-two inches long, which were marked by wooden buoys that took on square, wedge and round shapes. Many of the traps had old plaster laths nailed to oak frames.

These lobstermen spent winters in fish shacks, garages or basements building new traps and repairing old ones—dusty, dirty work. They knit trap heads at home during the evening. Rickety traps that had been weakened by storms and the wood-boring shipworms that honeycombed the wood were often rebuilt until their frames could no longer hold nails. Most lobstermen did not buy completed traps; instead they chose to make their own from either sawmill-cut oak trap kits or from their own milled pot stock.

Others, like the late Lane's Cove lobsterman Eino "Yiksi" Leino and Gloucester lobsterman Carl Fiers, even made homemade traps from materials gathered from the woods, shoreline and building demolition sites. Leino used oak saplings for both the framing bows and laths of his half-round traps. He made the laths by splitting saplings in half and his traps' side head rings out of boxwood and oak shoots. Leino used flat beach stones for trap weights. Fiers cut up oak pallets for his pot stock.

At the same time, around 1960, Dacron rope and nylon heading twines had begun replacing the rot-prone sisal heading twines and pot ropes. The sisal heads had to be replaced every season. The lobstermen had to make do with what was available. The late Paul Woodbury of Rockport, Massachusetts, created Dacron rope and the machinery to make it. He owned and operated the Rockport Twine and Rope Company at Rockport, Massachusetts. Woodbury's machinery spun rope out of the Dacron he reclaimed from rags.

The Harvesting of Lobsters and the Big Changes

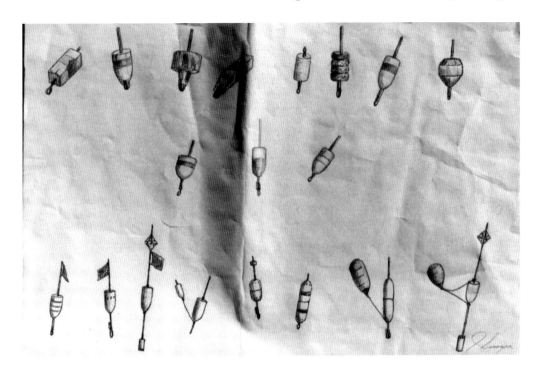

This Joe Sinagra drawing illustrates the old styles of wooden and synthetic buoys versus the synthetic buoys of today. Many of today's buoys have flags on their sticks.

By the early 1980s, the thirty-six-inch-long by twenty-three-inch-wide and fifteen-inch-high bear or wraparound trap entered the Cape Ann lobster fishery, quickly gaining widespread acceptance. The bear trap, which got its name for being longer, higher and wider than its predecessor, and for having machine-made nylon and polypropylene netting wrapped around its sides and ends, far out-fished the old-style lath trap. The bear traps' nettings were also cut out of the machine-made twine. These traps were heavy to handle, especially out of the water, were subject to ship worm damage and they moved easily during storms. To combat worm damage, some lobstermen soaked them in a foul-smelling, very toxic trap dip, while others used another new product on the line (vinyl-coated wire mesh) to close in their tops and bottoms.

Polypropylene ropes and PVC buoys and their special bonding paints showed up. Trap factories like Tucker Traps of Elliot, Maine, also popped up, not only cutting stock, but also making thousands of different sized completed traps, which they would even deliver right to the wharf.

About that same time, small numbers of the new all-wire trap began to filter into Cape Ann from Maine, thus beginning the wire trap revolution here. Many Maine lobstermen were getting rolls of this wire or trap kits and were building these during their off-seasons for themselves and also to sell on the side. Although the vinyl wire had been around since the late 1950s, it didn't catch on then.

James M. Knott Sr., president and CEO of Riverdale Mills Corp. of Northbridge, Massachusetts, has the distinction of putting "the very first, all-wire plastic-coated wire trap in the Atlantic Ocean…in 1957." While fishing a handful of heavy wooden traps in the summers by his Gloucester vacation home, Knott kept thinking, "There has to be a lighter and better trap than those heavy wooden traps. The first traps I built were nothing like those you can buy or build today. But, they worked."

His company is one of the current leading manufacturers of vinyl-coated wire mesh. Riverdale Mills brand Aquamesh prides itself on its unique Galvanized After Welding (GAW) process. A brand by Shepherd Lobster Wire in Newcastle, Maine (Rhino Tuff), a new Italian brand (Sea Plax) manufactured by Cavatorta and a brand new Chinese brand (Super Sea) are the other leading vinyl-coated wire brands. Beginning in 2002, Riverdale Mills also sells the trap wire, kits and the needed equipment, like wire cutters and clip guns, directly to lobstermen as well as to trap companies and gear supply outlets. Incidentally, those wire traps that Tim Sullivan made in 1988, and is still fishing in 2005, were made of Knott's Aquamesh wire.

With the rising costs of completed traps—sometimes as high as seventy dollars per trap—more lobstermen are buying kits and building their own traps, often saving up to 50 percent per trap this way.

"I found I was getting more lobsters per haul than I was with wood. Most of my traps stayed on the bottom during some pretty heavy weather; water flows through the wire trap better than through a wooden one. Those that didn't (if I could find them) were easily repaired," Knott says. Knott also discovered that the wire trap is lighter out of the water and heavier in the water than a wooden trap, "about five times as heavy as the wooden trap [in the water]."

But, Knott had a tough road ahead convincing the traditional wooden-trap lobstermen (who reasoned, "if it ain't broke, don't fix it") to switch to wire. Initially, "acceptance was slow; for many years sales were often less than a truckload a year," says Knott. While exhibiting at the first Fish Expo in Boston, an old lobsterman told him, "It looks good, but it ain't gonna catch no lobsta." After trying some of Knott's completed wire traps offshore in the deep water, the late fishing legend Bob Brown told him, "They don't work. I never caught a lobster with them." Little did Brown realize then, those traps would have probably worked if they had landed upright. Initially, lobstermen who fished wire traps in the deeper water had a balancing problem with them. The lobstermen and the industry corrected this by tying the traps' rope bridles to their upper corners and readjusting the traps' weights.

Incidentally, Knott's first customer was the Rockport Twine and Rope Company. While a few Cape Ann lobstermen then used that wire for the ends and sides of wooden traps, a handful of others, including the late Lane's Cove lobsterman Arthur Gaudreau, later made small-sized (under thirty inches long) complete wire traps for themselves. I can still remember other Lane's Cove lobstermen telling Gaudreau, "You're wasting your time with those traps."

The Harvesting of Lobsters and the Big Changes

But, Knott kept marching on, and by the late 1970s, J. Pike Bartlett Jr. of Maine, who later formed the Friendship Trap Company of Friendship, Maine, "really got it going; he opened the doors," says Knott. Bartlett was then instrumental in getting Maine lobstermen to use wire traps. Today, Friendship Trap is one of the largest wire trap manufacturers.

Friendship Trap Company, also having a branch at Columbia Falls, Maine, and an affiliation in Wakefield, Rhode Island, is "the largest in the business. We produce more custom-made traps than anyone else by far. There's no such thing as a stock trap anymore. We made one hundred thousand traps last year [2002]. We make literally hundreds of different styles; no two trap styles are identical," explains general manager Mike Wadsworth. He further added, "Annually, 85 percent of our customers [up and down the east coast, including many on Cape Ann] buy completed traps ready for the water. Probably the next 10 percent buy incomplete trap cages with the bricks, and the last 5 percent buy the kits," he added. The company, with its approximately forty-five employees, also sells used traps and lobstering supplies, delivering them right to the dock.

Today, Knott estimates that 80 percent of New England and 60 percent of Canadian lobstermen use wire traps, "and the number grows every year." Lobstermen have discovered the difference between fishing wire versus wood is like night and day, translating to more fishing time, fewer hang downs, less maintenance and storm damage, longer gear lifespan and ease of repairs and handling.

Wire traps today come in standard 30-inch, 36-inch, 42-inch and 48-inch lengths; widths vary from 18 to 24 inches, while heights range from 12 to 15 inches. Offshore lobstermen fish 50-inch-long traps. Every lobsterman believes in his mind that his trap works the best. The most popular mesh size and thickness is 1 $\frac{1}{2}$-inch by 1 $\frac{1}{2}$-inch by 12-inch gauge. The four-foot-long traps have double parlor heads. Most traps also come fitted with small-mesh shrimp twine nettings, often with rings in the parlor heads and the side heads with 6- to 7-inch-diameter rings. Lobstermen believe lobsters can crawl easier on the shrimp twine. The only drawback of the shrimp twine is that it offers more resistance to moving water, and those traps with the shrimp twine move easier in storms. Dick Winchester, owner of Winchester Fishing Gear Company in Gloucester, reports the 36-inch-long, 23-inch-wide, by 15-inch-high "Casco Bay" trap, which has similar dimensions to the old-style wooden bear trap, is his most popular model. More lobstermen in Maine are going to "super" traps that are 4 feet long, 24 inches wide and 15 inches high. These traps offer a lot more space to hold lobsters.

The wire lobster trap, along with its accessories, is still evolving during the never-ending quest for the perfect trap. Traps made from yellow-colored wire are one of the latest trends. Lobstermen believe these fish better in the deeper and darker water than the customary black and green-colored wire traps. Could the yellow be more visible to the lobsters? Traps can now come with long-lasting runners made of either concrete, which also act as weights, or polymer plastic. To guard against shipworm damage, most of the wooden runners come pressure treated or are cut from "jungle wood," resembling mahogany. The shipworms are repelled by this wood.

Hydraulic Pot Hauler

Lobstermen once hauled their traps either by hand or winch head. These physical and time-consuming techniques limited the string sizes and the ranges. Most of the hand hauling was done fishing single traps out of skiffs and dories in the shallow water and even from small wind-powered sloops before that. I remember how backbreaking the hand hauling was after storms, which wrapped kelp around the traps and their buoy lines.

Powerboats equipped with belt-driven revolving winch heads offered their owners greater fishing ranges and bigger strings, although the winch heads required constant vigilance and arduous repetitive hand-over-arm movements. A lobsterman always had to be on high alert and ready to kill the main engine if the rope on the winch head got twisted. Some operators' hands have been pulled into the winch heads and broken by such snarls.

The hydraulic pot hauler, which showed up along the east coast in the 1960s, slowly phased out the winch head and most hand hauling. Much safer, faster and more powerful than the winch head, the hydraulic pot hauler, which is powered either by the main engine or an auxiliary engine, has allowed boats to further increase their string sizes and also fishing ranges, especially in deep water. It's like having an extra crewman aboard. Instead of wrapping rope around a winch head, the lobsterman simply has to place the rope between a pair of motor-driven rotating plates and control the whole hauling process by opening and closing a control valve which regulates the flow of hydraulic fluid. The lobsterman can work on a just-hauled-aboard trap while the next one in line is being yanked in. Larger pot haulers, especially made for offshore lobster boats, can haul up single traps at a rate around 500 feet per minute and even pull up a maximum dead weight of approximately 2,500 pounds. Smaller versions, including electric pot haulers, are now made for skiffs.

"I was a textile man, and I came to Maine to manufacture lace. I wanted to become a lobsterman, but not one having to use a winch head," says Robert Crowe Sr., owner of Marine Hydraulic Engineering of Rockland, Maine. So Crowe, who describes himself as being "the first person to promote the hydraulic pot hauler on the east coast," created a smaller version of the already-existing large-size Marco Hydraulics hanging block hydraulic hauler. His reasonably priced hydraulic pot hauler, the Hydro-Slave, could be mounted either to a boat's bulkhead or to its side.

Crowe initially had trouble selling his pot hauler. After installing one in a demonstration boat, he stopped at ports from Kittery, Maine, to Grand Manan. "No one bought it," he says. Finally in 1964, "I sold the first one to Crehaven, Maine, lobsterman Donny Simpson. The other lobstermen laughed at him." Simpson fished trawls, and "soon he doubled his lobster catch. It took about another year before others got interested; they didn't trust the new hauling system," adds Crowe. The pot hauler slowly gained acceptance. Today, lobstermen wouldn't dream of not having their boats equipped with one of them. His Hydro-Slave brand and a Canadian version made by Hydraulic Hauling Systems are both available today.

The Harvesting of Lobsters and the Big Changes

Friendship Trap Company, headquartered at Friendship, Maine, is one of the largest wire trap manufacturers today. Here, potential customers wander through its booth at the Maine Fishermen's Forum

THE FIBERGLASS LOBSTER BOAT

Great changes have occurred in lobster craft since the industry's beginning in the 1800s. Lobstermen have gone from wooden, sail-powered boats (especially Friendship sloops, hand-rowed dories and peapods, small double-ended skiffs, gas engine-propelled inboard and outboard skiffs, dories, and open and closed-in power boats) to larger, faster, mainly fiberglass, diesel-powered Canadian and American-made work boats to work their gear. Lobstermen can now choose from a selection of about fifteen foreign and domestic name brand diesels, which can range in horsepower from one hundred to over one thousand. Outboards have come a long way, too. The new ones have to meet the federal government's Tier-Two emissions standards. The modern four-stroke outboards are very fuel-efficient and reliable.

The late Ephraim Atkinson from Cape Sable Island, Nova Scotia, is credited with making the first Cape Islander-style Canadian lobster boat in the early 1900s. Atkinson is the uncle to the current well-known boat builder Bruce Atkinson, owner of Bruce Atkinson Boatbuilding of Clark's Harbour, Cape Sable Island, Nova Scotia. Many U.S. and Canadian lobstermen today use the high-bowed and beamy Cape Island and the narrower, shallower and lower-bowed Northumberland Strait-style Canadian boats, as well as a number of Massachusetts, New Hampshire and primarily Maine-style lobster boats.

Here, Mr. Parsons is hauling lobster traps off Salt Island in Gloucester, Massachusetts, probably in the early 1900s. *Courtesy Martha Hale Harvey, Cape Ann Historical Association.*

Incidentally, Harold Burnham of Essex, Massachusetts, who operates H. Burnham Boat Yard and sloop boat charters, specializes in building historic boats. He also takes out customers on his sail-powered *Chrissy*, a small Friendship sloop, in the summer to lobster the old-fashioned way in Ipswich Bay. To haul a trap, Burnham nearly stops his *Chrissy* the way the old-timers did by turning the forward sail hard over one way and the stern rudder hard over in the opposite direction.

The last big lobster industry revolution, which took off during the late 1970s, has been the fiberglass composite lobster boat. Not surprisingly, many lobstermen also greeted the "glass" boat with great resistance, some ridiculing these "Clorox jugs." Others said, "God made trees out of wood, not fiberglass." A few old-time wood boat traditionalists even said they "wouldn't step foot on a fiberglass boat for fear of it rolling over." The traditionalists also wondered how the fiberglass would cure over the years, and if it would hold up. Fiberglass has become the hull material of choice for lobster boats commonly up to fifty-five feet long in the United States and Atlantic Canada.

Pre-1970, wood was *the* small-boat-building material. Even though wooden boats have a special warmth to their appearance, are sea kindly to their owners, are heavy in the water and can be rebuilt over and over, they are slower in the water, require annual painting to prevent bottom worm damage and top side fresh water rot and they

The Harvesting of Lobsters and the Big Changes

These open "Strait"-style Canadian-built lobster boats were used by Boston Harbor lobstermen in the mid-twentieth century. *Courtesy Bill Atwood, William Atwood Lobster Company.*

have a greater risk of leaks. Many thirty-year-old or older wooden boats are still actively lobstering today. A few lobstermen still prefer wooden boats.

Quality wood, especially oak, used to be readily available at fair prices. Many boats' keels and framing were oak, while their planking was cedar, fastened with either galvanized nails, or even better, silicon bronze screws. Lobstermen frequently relied upon local boat builders—often lobstermen themselves who built boats during their off-seasons—for their boats. Skilled wooden boat builders have become a dying breed.

In the early 1960s, Webber's Cove Boat Yard of East Blue Hill, Maine, made either one of the first or the first fiberglass composite lobster boat on the east coast, which happened to be one of their thirty-four-foot-long Webber's Cove models. The boatyard had been manufacturing fiberglass launches for the U.S. Navy. Webber's Cove boatyard simply went one step further and made a lobster boat from the same process.

"I guess we have since made around two thousand lobster boats, including close to five hundred of the thirty-four-footers," says third-generation boat builder Matt Cousins. He and his brother David now own and operate Webber's Cove Boat Yard. His father "Jug" and grandfather Cy started the business. Incidentally, Webber's Cove made my current lobster boat, a South Shore 20 in 2004. This is a Calvin Beal Jr., design.

The fiberglass fibers, which are woven into sheets of matting and cloth, are actually made from melted glass. The composite is formed by mixing a gooey liquid resin or plastic polymer and a hardener with the matting and cloth. Once hardened, this combination forms a strong, impervious layer. Layer upon layer of such laminated fiberglass is laid upon different molds, many based upon traditional Maine and Atlantic Canadian lobster boat models. Upon customer request, some boat builders use synthetic or natural cores between layers of fiberglass to increase hull thickness while reducing the hull weight.

But "Solid [no core in between the fiberglass layers] is standard today, and coring is optional. Producing a glass boat is a lot easier than wood, but the laminating is a bit fussy; you have to look out for a few dos and don'ts," says Wayne Beal, who owns and runs Wayne Beal's boat shop in Jonesport, Maine. Beal manufactures twenty-eight-foot, thirty-four-foot, thirty-six-foot and forty-foot-long lobster boats. The fiberglass on his thirty-six-foot-long boats is two inches thick at the keel, three-quarters of an inch thick on the bottom and a half-inch thick on the sides. It takes Beal about four to five weeks to laminate a hull and top and about three months to do a boat completely.

Further east, the majority of Wedgeport Boats Ltd.'s "Novi-style" thirty-five- to forty-five-footers have "a solid skin and not a sandwich construction. We turn out a hull every twenty-six days. From start to finish and ready to go fishing takes between eight to twelve weeks," explains Harland Martell, second-generation boat builder, owner and president of Wedgeport Boats Ltd., Lower Wedgeport, Nova Scotia. Incidentally, the company made their last wooden boat in 1982.

Helped by wooden boats' escalating building and maintenance costs and the trend of more lobstermen fishing longer seasons further offshore and the need to get there faster and to carry more gear, the fiberglass composite lobster boat slowly caught on by the 1970s. The fiberglass composite boat provided a superior alternative to wood, offering a seamless, strong, rot-free hull requiring less maintenance and offering easier repairs and greater speed. These boats still have to have their bottoms painted annually to prevent seaweed, barnacle and mussel fouling, and their hull colors occasionally polished or repainted. Their hull zincs, which prevent electrolysis, have to be replaced just about every year, too. Repairs in glass boats can be easily made, often just requiring sanding, filling in and then glassing over. Like wood, glass boats also have a long life span. Many of these current working lobster boats were built in the 1970s.

Seeing the handwriting on the wall, many wooden boat builders converted to making glass boats, either as completed boats, kits, or just hulls. New fiberglass boatyards also began popping up, including the former Bruno and Stillman boat yard of Newington, New Hampshire. Many of their original thirty-five-foot, forty-two-foot and later fifty-five-foot models still work the east coast today. Downeast Maine and Nova Scotia dominate lobster boat building today. The Beals, Libbys, Cousins, Youngs, Hollands, Lowells, LeBlancs, Greenwoods and Atkinsons are big builder names and, in many cases, designer names, too.

"About 65 percent of the current lobster boats in the U.S. are Carroll and Royal Lowell designs. The brothers started in 1971. They are responsible for the hard-chine hull revolution. Most small and medium-sized working glass boats in harbors today,

The Harvesting of Lobsters and the Big Changes

Most lobstermen today order completed traps from trap companies, which will deliver orders right to the dock in tractor-trailer trucks.

including the forty-nine-foot and fifty-six-foot Duffys, fifty-five-foot Bruno Stillmans, the DMRs, Jarvis Newmans and JC Boats, are Lowell designs. They were designed as lobster boats," say grandsons Jamie and Joe Lowell. They run Lowell Brothers at Yarmouth, Maine. They also design boats and build the Lowell forty-three-foot fishing boat. Royal Lowell passed away in 1983.

Currently, the fiberglass composite is still the material of choice, but the lobster boats' sizes, power plants and accessories are changing. Wayne Beal, like many of the other fiberglass boat builders, has noticed that lobstermen nowadays "want to have as big a boat as they can get, with the top of the line equipment, including bigger pot haulers and the best of the electronics, and a speed of at least twenty knots. They take no short cuts." At Holland's Boat Shop of Belfast, Maine, Glen Holland has even installed a 1,300 horsepower Mac diesel in one of his new thirty-eight-foot boats. There almost seems to be a rat race in some of the coves and harbors to have the biggest and fastest lobster boat. Most lobster boats today are in the thirty- to fifty-foot-long range, some having price tags well over $100,000, even surpassing $400,000. Like the old redfish fleet out of Gloucester in the 1950s and 1960s, many of the lobster boats have only gotten bigger and fish further from port. Hopefully, the lobster fleet won't disappear like the redfish fleet.

VI
Lobster Marketing

Ever since the early Puritan settlers first learned from the Indians how to utilize the lobster, it has been one of the most prized articles of food in the New England states. At first, as most settlers lived on or near the coast, each family could easily secure its own supply, but as the settlements gradually extended further inland, this became inconvenient, and soon it became customary for certain persons living on the coast to supply the wants of the inland settlers, and thus the commercial fishery was established.

Bulletin of the United States Fish Commission, Vol. XIX

The Changes

After that birth, the world has slowly discovered the American lobster and it has found a solid niche in the world seafood market. Thanks to the tremendous changes in lobster marketing since the late 1970s and globalization in the twenty-first century, which has helped introduce the lobster into new cultures, people can now get the cold-water lobster year-round. Lobster is available for purchase whole or in parts, either "red" (cooked) or "green" (uncooked). Markets can now handle huge volumes of annual U.S. and Canadian Atlantic coast lobster catches. Improved lobster holding, shipping and processing technologies, rapid transportation (now also by air freight) and communications have also evolved.

In the 1800s to the mid-1900s, canned meat and whole lobsters were the primary consumer end products that fueled the industry. Then keeping the fishery going partly depended upon transporting lobsters from the lobstermen and local dealers to centralized canneries and larger dealers. The *Bulletin* describes this history of the cannery.

> *When the canneries [Maine] went into operation, they usually worked during the spring, early summer and fall, and they furnished a ready market for all the lobsters that could be caught. As at certain places on the coast, the canneries were the only market for lobsters.*

Lobster Marketing

The fishery would cease as soon as the canneries stopped [for the season]. The canning industry in the states practically ceased to exist in 1895 and since then the whole catch has had to be marketed in a live or boiled condition.

Lobsters were transported over the land by rail and later truck and over the water by specialized vessels called "lobster smacks." "Dry smacks" carried iced-down lobsters held in either crates or barrels in their holds, while the "wet smacks" kept their cargo alive in flooded holds with circulating seawater. These dealer and independently owned trailer trucks of the sea routinely transported United States and Canadian lobsters to east coast markets. The *Bulletin* writes,

As early as 1830, smacks from Boston and Connecticut visited Harpswell [Maine] for fresh lobsters…

The smack fleet had been gradually increasing as the live lobster trade extended, and by the time the canneries closed permanently, they had extended their vessels to every point where lobsters could be had in any number.

Lorraine Reinhardt, a seventy-five-year-old semi-retired lobsterman from Le Havre, Nova Scotia, who crewed on a 120-foot dry smack owned by O.K. Services of Le Havre in 1942, once transported Newfoundland-caught lobsters—approximately four hundred crates each trip (four thousand pounds)—to the former Consolidated Lobster Company Ltd. at Bay View, Massachusetts, about a three-day journey each way. Bay View, which is part of Cape Ann, abuts Ipswich Bay. This lobster company was the largest in the world that used to can lots of lobster meat. Consolidated Lobster Company Ltd. closed in the early 1960s, and the University of Massachusetts now operates a marine station there.

"About nine-tenths of the lobsters caught in Maine waters are shipped in the live state. The principal shipping centers are Portland, Rockland and Eastport, which have good railroad and steamship facilities with points outside of the State." Lobsters were shipped over the road in wooden sugar and flour barrels with holes drilled in their bottoms, approximately 140-180 pounds per barrel, each capped with ice, seaweed and damp burlap or sacking. "Packed in this way, lobsters have easily survived a trip as far west as St. Louis."

In addition, the *Bulletin* describes one of the first attempts to ship lobsters overseas in 1877: "Owing to the high prices realized in England for live lobsters, attempts have been made to ship live American lobsters to that market, generally from Canadian ports. Messrs. John Marston and Sons of Portland [Maine] made a trial shipment of 250. They were placed [on board on deck] in a large tank constantly supplied with fresh sea water… This trip was fairly successful as only 50 died." Little did these lobster dealers then know that overseas shipping would become common in the future, but by air instead of water.

Naturally, during this time, seasonal lobster shack type restaurants and inns and hotels of a growing tourist trade along much of the northeast coast also created markets for whole, live and cooked lobsters and their meat. When I started lobstering in 1960, many

Ipswich Shellfish Company, Ipswich, Massachusetts, has the largest indoor trickle-down, or spray, lobster holding system in the world. The system can hold over 500,000 pounds of lobster. Here, company officials (*left to right*) Mike Trupiano, Alexis Pappas (also the company's vice president) and Fred Fullerton pose in the lobster holding area as water trickles down from overhead piping into totes holding lobsters.

lobstermen on Cape Ann sold to such restaurants, which often opened on Memorial Day and closed after Labor Day. Once these markets closed for the season, the lobstermen had to either find another dealer, stop fishing or truck their lobsters into Boston, the big lobster distribution center.

Spiros G. Tourkakis, executive vice president of East Coast Seafood explains that the U.S. lobster market up to the late 1970s "was very regional, especially within the New England area, also reaching out to New York and Philadelphia. Boston was the big [lobster] center. The products were primarily live lobsters and fresh and frozen lobster meat [the latter often a product of Canada packed in 11.3-ounce cans]. Everything was distributed by truck."

Boston lobster dealers like Atwood Lobster Company, Bay State Lobster Company, Hines and Smart Lobster Company, Mayflower Lobster Company, James Hook, Willard-Daggett Company and the Consolidated Lobster Company Ltd. not only bought and distributed local product, but also huge volumes of Maine and Canadian lobsters that were trucked there. These dealers controlled the markets and boat prices.

As a teenager back in the 1940s, third-generation lobster dealer William Atwood offloaded freight cars in Boston filled with barrels of Maine lobsters (each barrel holding fifty or seventy-five pounds) bound for his father's Atwood Lobster Company. Atwood also has memories of re-loading barrels of his father's lobsters onto freight cars bound for major U.S. cities where customers were waiting for these orders. He now owns the William Atwood Lobster Company of Spruce Head Island, Maine, one of the state's largest and best-known lobster companies.

With the building of the interstate highway system in the 1950s, refrigerated tractor-trailer trucks became the main transporters of live lobsters in the United States and Canada. These lobsters were originally shipped in heavy rectangular-shaped wooden crates holding about one hundred pounds each, and later in easy-to-handle, lightweight plastic crates with lids that snap shut.

Today, the big trucks can transport up to forty thousand pounds of lobsters per load in their forty-three-foot-long bodies. A few smacks still ferry lobsters from the islands off Maine to the mainland.

Early Marketing Problems

Prior to the late 1970s, lobster buyers and marketers were sometimes plagued by high summer and fall lobster mortalities, which often occurred when gluts from the Fourth of July and fall shedder runs were jammed into dealers' holding tanks. The inability to suddenly move large volumes of lobsters also contributed to this problem. In contrast, lobster buyers and marketers also experienced seasonal live lobster shortages during the winter and spring. Most of the indoor lobster holding systems at the time merely re-circulated warm, low-oxygen harbor water through the tanks. That and the crowding further stressed these summer and fall lobsters, which were already weakened by molting.

Mounds of dead and dying lobsters—"weaks" plucked from the tanks—sometimes littered the tanks' passageways. Workers at some companies went through these lobsters and often broke any questionable lobsters' tails apart from their bodies and sniffed these for odor. Lobsters that smelled were discarded, while the others were cooked, and their meat was salvaged.

The boat prices during gluts went down, down, down. Cheap prices helped diminish some of the dealers' losses. Supermarket sales occasionally helped move large numbers of these lobsters. As last resorts, dealers occasionally either shut off their lobstermen until they could empty their holding tanks or floated the crated surplus catches in harbors and coves. They had to largely depend upon coastal tidal lobster pounds for winter and early spring supplies, too.

Correcting Old Problems

Better lobster holding and shipping equipment, followed by broader markets, including lobster processors, have largely eradicated those mortality and glut problems along with seasonal shortages of lobsters. Tourkakis further explains, "The 1978 deregulation of the airline industry allowed air freight, and at the same time, the technology of holding lobsters [Re-circulating Aqua Systems—RAS] improved. Suddenly, the market is national and soon to be worldwide. We made our first shipment to Europe in 1985. Now 90 percent of the live lobster business goes overseas," says Tourkakis. Europe and Asia are big lobster-eating continents. Clearwater Lobster Company, Ltd., of Bedford, Nova Scotia, credits itself with "opening up the European and Asian markets," says co-owner Colin McDonald.

Nowadays, most lobster orders are airfreighted out in special aluminum cargo containers, LD-3s, usually 2,400 pounds per unit. Graded "shippable lobsters," the strongest ones with firm to hard shells and each having two claws, within these industry standards: chix (1–1.24 pounds), quarters (1.25–1.49 pounds), halves (1.5–1.74 pounds), selects (1.75–2.49 pounds), jumbos (2.5 pounds and up). Culls, lobsters with only one claw, are also shipped sometimes and broken down into these industry standards: small culls (1.00–2.00 pounds) and large culls (2.01 pounds and up) are first packed in standard 11- to 50-pound cardboard shipping boxes. In transit, the boxed lobsters are kept cold by individual gel packs and sometimes damp by wet newspapers or seaweed. The boxes often bear the dealer's name and the grade of lobsters. Most of the LD-3s are lined with reflective foil to further protect the perishable contents from the cold and the heat. These packed lobsters will usually "live two days out of the water. Shipping mortality often averages around 2%," explains Mortillaro Lobster LLC owner Vince Mortillaro. This Gloucester company annually ships out about 5 million pounds of lobsters. Most dealers quote prices to their customers on Fridays and ship them product on Sundays and Wednesdays.

East Coast Seafood—Leader in the Industry

East Coast Seafood's parent company, American Holdco Inc. (AHI), whose "goal is to become the world's most reliable vertically integrated seafood supplier with a focus on lobster," blossomed out of the former East Coast Lobster Pool off Route 1 in Peabody, Massachusetts. "Mike [Tourkistas, president and CEO] took over East Coast Lobster Pool from his uncle in 1981," explains executive-vice president Spiros Tourkakis, who also started the Over-40 Boston Braves Soccer Squad in 2001. He is a goaltender for the team. Both Mike and Spiros emigrated to the U.S. from Ellinikon, Greece, a small suburb of Athens—the former in 1971 and the latter in February of 1978—ironically during the blizzard of '78. They are legends in the lobster industry today.

Tourkakis is also credited with "helping to convince us [IPL Products Ltd.] to develop the design of the Flap Nest plastic lobster crate in the spring of 1999. As of 2004, we have made in excess four hundred thousand of these," says Eric Fredrickson, marketing manager for IPL Products Ltd., based in Worcester, Massachusetts, and also St. Damien, Quebec. IPL makes plastic packaging containers for the seafood industry, including the popular plastic lobster crates.

East Coast grew from handling approximately two hundred thousand pounds of lobsters its first year to approximately twenty-four million pounds in 2000. AHI is headquartered

Lobster Marketing

Mike T. Caudle, "the Tank Doctor" and owner of the Tank Doctor Product Systems, poses at The Lobster Outlet, York, Maine, while installing one of his revolutionary trickle-down, or spray, lobster holding systems. Caudle resides at Enfield, Nova Scotia.

in Lynn, Massachusetts. AHI's holdings also include processing plants and indoor and tidal lobster pounds in Atlantic Maritimes, Worldwide Perishables, Seafood Procurement and Marketing and East Coast Europa, which has offices in Paris, Madrid, Milan, Brussels and London. East Coast Seafood is "the leading supplier of live lobsters to Europe's largest food retailers. East Coast services customers around the world, insuring you always receive top-quality product and services 365 days a year," says its information brochure.

Seafood Procurement and Marketing buys lobsters "from New York to Newfoundland," while Worldwide Perishables trucks lobsters and makes the air freight arrangements. "We also have our buying stations—contract buyers or alliances along the coast who either help with the supply or pack it or both. We ship directly out of these packing houses," says Tourkakis.

"Our primary focus has been to create new markets and also to find new products. Our business is changing. We are preparing for the future by investing in technology to streamline our operations and to reduce our accounting and administration, sales, transportation and logistics costs," explains Tourkakis.

Evolution of Lobster Holding Systems

The primarily open, unchilled, stacked, wooden and later fiberglass tank indoor lobster holding systems have lately evolved to indoor, closed, chilled, cement tank systems or

East Coast Seafood officials (*left to right*) Spiros Tourkakis (vice president) and Mike Tourkistas (president) pose at their company's headquarters at Lynn, Massachusetts. Both men are Greek immigrant success stories. East Coast Seafood is the largest lobster wholesaler in the world, annually marketing over twenty million pounds of lobster.

indoor, closed, chilled, trickle-down or spray systems. The latter two systems have also helped maintain a year-round supply of hard-shelled lobsters during both the summer when new-shellers are prevalent and the winter and early spring when newly caught lobster landings usually plummet.

Story has it that Mr. Guiseppe Brandano of Haverhill, Massachusetts, invented one of the first closed re-circulating water tank chill systems, "the Brandano system," during the late 1930s. These systems had costs: the toxic buildup of ammonium from lobster wastes and of heavy metals from the corrosion of the chilling coils killed lobsters. The water had to be changed and re-chilled every two to three days. Most of Brandano's customers made seawater by adding synthetic salts to tap water.

Those problems have been largely eliminated. "I built the first re-circulating aqua system for Spiros [Tourkakis] and Mike [Tourkistas] in 1983. I put a bio filter into their old holding tank and replaced the chilling coils with new ones made from non-corrosive titanium," recalls Tom Lauttenbach, research and development department director of Marine Biotech of Beverly, Massachusetts.

These open and closed re-circulating aqua systems can not only chill the tank water to just about freezing (twenty-nine degrees Fahrenheit) year-round, but also reduce annual lobster shipping and holding mortality to below 5 percent by lowering the lobster's metabolism. Dealers now "season" their lobsters several days before shipping them. RAS allows lobsters to be both shipped long distances and to be kept alive anywhere. After grading their supplies, dealers either float their lobsters (one hundred pounds per crate) or let their lobsters roam free in tanks.

Four major processes occur in Marine Biotech's aqua systems. First, the temperature is controlled by chillers and condensing units. Mechanical filtration separates the solids. Next there is the gas exchange, or the addition of oxygen and removal of carbon dioxide, often simply accomplished in small systems by the fall of water from one tank into another. Finally there is the act of biofiltration: beneficial bacteria, living within the tanks' submerged gravel beds, break down toxic ammonium into less poisonous nitrates. The gravel acts as substrate for the bacteria to settle on.

As word spread of this innovation and its installation at East Coast Seafood, Marine Biotech soon got calls from other wholesalers who "had their sights on the newly-opened European trade, to build systems for them and also their second partners who received the overseas shipments during the 1980s and early 1990s," says Lauttenbach. "We have installed over three hundred systems in the retail and commercial [lobster market] sectors in this country and abroad, even as far as South Africa," says company president Steve Aldrich. His company is one of several which install such re-circulating systems. Marine Biotech also services the aquaculture industry.

The latest advancement in lobster holding systems has been the trickle-down or spray system. "Six to seven years ago, I developed the spray system. I played with it over ten years. It took a while to make it right. The density is the big thing with the spray systems. You can keep the most amount of lobster in the least square footage. The ratio of lobster to water is three to one with the spray system. The opposite is true for the cement tank system," explains Mike Caudle, the energetic and enthusiastic, largely self-taught forty-six-year-old "tank doctor," owner and founder of the Tank Doctor Product Systems based at Enfield, Nova Scotia. He has been a forerunner in the recent lobster holding system revolution.

"I started building tank systems at Kitchener, Ontario, in 1986. My first system was for my brother who had a fish market–wholesale outlet there," explains Caudle, a former auto mechanic who grew up in Halifax.

"I began making one hundred-gallon aquariums and then progressed to two thousand-pound [lobster holding] wholesale systems, and lastly to ocean systems in Nova Scotia," he says. Many of Caudle's closed, cement tank lobster holding systems, where lobsters are often floated in crates, are still in wide use in the lobster industry today. But Caudle just knew there was a better way to inventory lobsters indoors and do away with the bulky cement tanks.

With his spray system, lobsters are stacked in totes or trays, usually sixty pounds to a tote, in a storage room. Often the totes are color-coded. Each color signifies a specific grade of lobster, which makes assessing inventory easy. Seawater originating from an onsite reservoir

Lobsters milling about in a dealer's holding tank wait to be packaged and shipped out. Live lobsters today are shipped all over the world.

constantly trickles down on the lobsters from valved overhead PVC piping, keeping them wet and cold, while supplying oxygen and removing wastes. Floor traps catch this water and it is re-circulated through a series of external adjacent stations for purification by beneficial bacterium living on pellet and honeycomb-shaped substrate made from plastic and by ozone sanitation. The water is then aerated and, lastly, chilled.

Caudle describes this process in-depth.

> *There is no new water. This system is completely closed. The pumps run continuously. The storage room doesn't necessarily have to be refrigerated, but it has to be at least insulated. The ozone treatment stops any disease the lobsters might have brought with them. The spray systems generally aerate themselves as the water leaves the piping and trickles down over the lobsters.*
>
> *These systems are a great improvement over the cement floating crate tanks. You can hold lobsters as long as six months in trays or totes in a temperature around thirty-six degrees Fahrenheit. You can even briefly bring the water temperature up and feed the lobsters.* [As an added bonus, the lobsters also stay clean in the totes.]
>
> *These systems have really caught on. The first big system I made was for Canadian Gold at Jeddore, Nova Scotia. This unit holds around two hundred and fifty thousand pounds of lobsters, just in totes.*

Today, Caudle's spray systems in Nova Scotia can hold about two million pounds of lobsters indoors.

In 2004 over an eight-month period, he installed what is, according to Caudle, the world's largest indoor lobster holding system at Ipswich Shellfish Company in Ipswich, Massachusetts. The tank is housed in a 150-foot-long by 30-foot-wide by 10-foot-high walk-in cooler, which formerly housed its floating tank, which could hold 70,000 pounds of lobsters. This spray unit, which is computerized, can hold 500,000 pounds. "We had over 400,000 pounds in there just before Christmas. The totes replace the tanks here. Now you have more lobster than water in this system," says Fred Fullerton, the general manager of Ipswich Shellfish Company's lobster division.

Chilled and purified seawater, which was trucked to the site from the system's three hundred thousand-gallon reserve, trickles down on the lobsters from overhead pipes at a rate of about six thousand gallons per minute, keeping them wet and cold while supplying oxygen and removing wastes. The lobsters have been graded and stored in color-coded, dated and tagged standard plastic totes. There are approximately sixty pounds per tote, stacked seven high in two parallel rows per pallet. This water is so bubbly and full of oxygen that Ipswich Shellfish Company calls their spray system "the champagne system." Twenty-four electric pumps re-circulate the water in the system. A backup generator is ready to kick in if the power fails.

Caudle's systems are also at work in Belgium, Scotland and England. Caudle and his assistants, Fred Budge and Colin McDonald, both from Cape Breton, Nova Scotia, are currently installing a twenty-five thousand-gallon system which can hold seventy-five thousand pounds of lobster for the Maine Lobster Outlet at York, Maine. "This will be the first complete spray system in Maine," says Caudle.

Each year the spigot flow of newly caught lobsters, especially in the United States, generally drops by mid-winter. Cold and stormy weather and low water temperatures help close that spigot. Dealers now use the indoor closed and chilled cement tank and trickle-down systems, along with further specialized techniques, to overcome this problem by inventorying lobsters for long periods of time. Atlantic Canada has led the way within the last two decades with indoor closed system lobster holding, which affords better climate control and reduces some of the risks of the traditional tidal lobster pounding, including disease, lobster nibbling and seaweed growth on lobster extremities.

"We keep the lobsters in pristine shape until we sell them, reduce the risks [of pounding lobsters], but have huge costs. We buy [and also pound] in December and sell from January through May. We also buy again [and also pound] in May and June and sell these through December," says Clearwater Lobster Company co-owner Colin McDonald.

This Bedford, Nova Scotian company, which has been "shipping live lobsters worldwide for twenty-six years" handles about ten million pounds annually. The company is supplied by both its own boats and other dealers.

Lobsters are placed individually in trays in waist-high water at Clearwater's two indoor holding facilities. Layers of these trays are submerged in the very cold tank water,

This historic photo shows what a lobster dealership looked like in the mid-1900s. The photo shows (*left to right*) "Mr. Zanti, Marsden F. Atwood (my father), William D. Atwood (my grandfather), Henry Wladkowski (in the tank) and Frank Zanti with the basket of boiled lobster. They shipped all over Boston, including Scollay Square and to most of the First National stores in New England," explains William Atwood. This historic photo, which also shows wooden barrels that were used to transport lobsters during that period, was taken inside the Atwood Lobster Company in Boston in the mid-twentieth century. *Courtesy Bill Atwood, William Atwood Lobster Company.*

sometimes at twenty-nine degrees Fahrenheit. The lobsters will feed when the food is practically placed right on top of them. "We often hold the lobsters three to five months before selling them," says McDonald.

Similarly, R.I. Smith Co. of Shags Harbor, Nova Scotia, has been tubing up to four hundred thousand pounds of lobsters annually since the late 1990s in its two indoor pounds, which are about forty inches deep. One to two lobsters (approximately two and a half to three pounds combined weight) are placed into eighteen- to twenty-inch-long by four-inch-diameter PVC tubes and then thirty-two of these tubes (about eighty pounds of lobsters), are stationed horizontally in separate wire cages, which are submerged in the big indoor pounds. The lobsters taken out of these tubes months later are "just about comparable to a new-caught," says owner Elnathan "Babe" Smith, who has over forty years of experience in the trade. The company exports its lobsters to the United States and overseas.

During the early 2000s, East Coast Seafood has even developed "the habitat packaging solution," which the company applies to its lobsters stored in a trickle-down system in Canada. Here, explains East Coast's brochure, *live lobsters are placed in plastic sleeves with individual compartments isolating each lobster throughout the holding and shipping process. The sleeves are easily removed from standard fish totes for packaging in our customized wax box designed to hold four habitat sleeves. The Habitat Packaging Solution not only reduces stress and damage on each individual lobster, but also reduces product handling further, maintaining the quality of the product.*

A number of Canadian dealers like R.I. Smith and Clearwater pre-test their potential pounding stock's blood for serum protein levels by extracting tail section blood with a hypodermic needle and getting a refractometer reading on it. The higher the refractometer reading—up to thirteen—the healthier and higher quality the lobster. Such lobsters are often so full of meat that they will literally explode when their shells are cracked after cooking.

Traditional Tidal Lobster Pounding

Many of the Maine and Atlantic Canadian tidal lobster pounds were built in the 1800s and have been the main source of supplying lobsters during the low-supply cold-water months.

Although Canadian and Maine outdoor tidal pounds have long helped supply the lobster market during the cold-water months, these pounds often posed greater risks to both the stored lobsters and the owners than the indoor systems. In theory, the tidal pounds first buy cheap in the fall and then temporarily sit on thousands of pounds of lobsters, which are usually sold for a profit, from January through March. Profits then can be several dollars per pound, often hitting six to seven dollars per pound by early April before dropping sometimes a dollar per pound at a time as Canadian fishing districts open and their new-caughts reach the market. The lobster pounds have also lately increased demand for lobsters in the fall and have kept up the boat prices then. They provide a good outlet for surplus catches.

"The lobster pounding business is worse than the stock market," explains Richard Carver. Carver and his son Albert are old pros who own three pounds having a total lobster capacity of about two hundred and fifty thousand pounds at Beals Island, Maine. Albert, the fifth-generation Carver to run lobster pounds, also manages Deep Cove Pound, one of the island's oldest pounds, built by his great-grandfather, the legendary Charles Henry Beal. Year after year the Carvers, like the other pound owners, including independent operators and lobster dealers, not only have to gamble with nature—especially weather and the lobster itself—but also with world markets, while hoping to come out ahead. They have all taken some hits over the years.

Beals Island has eleven lobster pounds, representing approximately "24 percent of Maine's lobster storage. 90 percent of the pounds are from Stonington east," says Albert. The ocean water stays cold much of the year here, rarely rising above sixty degrees Fahrenheit in the summer. Mid-coast Maine also has several lobster pounds.

Many tidal lobster pounds were built in the 1800s by damming off the fronts and sometimes the outer rims of natural coves and shoreline indentations usually free of fresh water runoff. Shoreline property was cheap and available then. The solid rims and fronts of these natural holding tanks run from the bottom to within two feet below mean high water to allow circulation at high tide and still hold water at low tide. Some of these rims and fronts are sometimes made from poured concrete, stacked up and bolted together cement T-blocks, or six-foot-thick walls of mud and clay sandwiched between planking, and their tops are further fenced by regularly-spaced boards which not only allow tidal flushing, but also prevent seaweed and flotsam from entering and lobsters from departing. In addition, each pound has one or two screened seaward-facing dams fitted with chutes or clapper doors. These chutes can be pulled to drain the pound.

The lobsters stored in pounds with muddy bottoms (versus those with rocky bottoms) have greater demand because those lobsters remain relatively free of undesired seaweed growth while in storage, especially on their antennae. "The water in the muddy bottom pounds stays cloudy, blocking sunlight necessary for plant growth," explains Richard Carver. The Carvers' pounds have muddy bottoms.

The Carvers' gambling usually begins in the fall as they fill their pounds with reasonably priced and healthy crate-run (as they come off the boat) lobsters bought from local dealers. They quickly separate and sell off the one-clawed, no-clawed and weak lobsters. "The stressed lobsters [the weak ones] will be eaten by the non-stressed ones; a good lobster will hum [vibrate] in your hand," explains Richard Carver.

Packing the lobsters into pounds further stresses the crustaceans, sometimes making them vulnerable to diseases, suffocation and cannibalism. The Carvers' store about two hundred and fifty thousand pounds of lobsters in their pounds. They feed, aerate and observe the lobsters to keep losses to a minimum. Besides adding air to the water, the aeration prevents the pounds from freezing and keeps the water stirred up and cloudy. The Carvers regularly feed their lobsters with salted cod racks trucked from Canada and medicated feeds which reduce cannibalism and ward off lobster diseases, especially red tail, a bacterial blood disease. Ironically, extremely cold water is also believed to pave the way to shell disease by shutting down the lobster's immune system and allowing the shell-dissolving bacterium to flourish. "If there is a disease problem, we will empty the pounds out, salvage what we can and sell to the processor, and if time permits, start over again," says Albert. Despite these preventative measures, by season's end, even during a good year, a 10 percent mortality rate is not uncommon for lobster pounding.

The Carvers usually begin emptying some of their lobsters "on a rising price" around February first and finish by the start of the annual Boston Seafood Show in March. They initially tow a lightweight scallop dredge behind a skiff to harvest their pounded lobsters before completely draining their pounds to get the rest. Divers help empty other pounds, too.

"One year, I waited, hoping for a $7 per pound price. Next, the price went down $3 in two days, and I had to sell for $3.50 per pound," recalls Richard. Their lobsters are trucked away to customers in southern Maine, Boston and New York. "Every pound owner has their own schedule," adds Albert.

Processors' Role

Canadian and Maine lobster processors have helped remedy some of the industry's past ills by providing a critical outlet for non-shippable lobsters, especially the weaks, the soft-shellers, the "bullets" or "ministers" (lobsters with no big claws) and lobster gluts. Processors help the industry supply customers with mainly frozen lobster and meat year-round.

"Not all lobsters are created equal. People [in the industry, especially the processors] have been experimenting with new lobster products and developing new markets for these, especially the lobster tail market," Tourkakis explains.

The percentage of non-shippable lobsters varies from area to area and season to season, sometimes making up 40 percent of the landings. Lobster quality is generally the poorest from July through December. The hard-bottom lobster is usually a better quality lobster compared to the soft-bottom lobster. The soft-shell, weaker lobsters seem to mass in the deeper, colder water on the soft bottoms in the summer and fall to get away from the inshore water that is too warm for them. Soft-shell lobsters congregate on the soft bottoms again in the winter to be in the now warmer water here. The hard-shell lobsters can also take more of a beating than the new-shell ones. During storms, lobsters don't get banged up as much living on soft bottoms as hard bottoms.

In the past, many lobster dealers cooked non-shippable lobsters for meat just to keep the lobsters in their tanks moving. This labor-intensive effort usually earned them little, if any, profit. Many dealers figure six to seven new-shelled chix lobsters are needed to make one pound of meat. They still do that to fill their orders for meat, but now ship any excesses to processors.

For the most part, the lobster processors have added stability to the summer and fall boat prices. The poor-quality lobsters used to deflate the boat price. Most processors often pay dealers around twenty-five cents per pound over the boat price for their non-shippable lobsters.

Canada has "about thirty processing plants. Many started thirty to forty years ago just canning lobster meat. This is a big thing; some process between five and fifteen million pounds of lobster every year," explains Jerry Amirault, chairman of the Maritime Lobster Processors Cooperative. Canada's processing season generally runs from May through Christmas, beginning with "canners" (lobsters which weigh under a pound) in May and June, followed by a lot of crated U.S. non-shippable product regularly trucked in during July through October, and ending with Canadian lobsters in December. It is estimated that Canadian processors often annually buy approximately forty million pounds of United States lobsters. Ironically, many of the crate-run Canadian lobsters trucked to United States dealers are trucked back again to Canadian processors after these have been graded in the United States and the shippables have been taken out.

Paturel International Company of Deer Island, New Brunswick, Canada, which belongs to AHI, is one of these processors. This company turns out "top-quality frozen lobster products," many of which have been individually quick-frozen and brine-frozen and sealed in vacuum bags under the Paturel brand. These include cocktail lobster claws, cooked frozen whole lobster, frozen lobster meat, raw lobster tails and frozen lobster popsicles.

Lately, the number of lobster processors in Maine has grown to approximately six, now including Cozy Harbor Seafoods at Portland, Maine, which is also a major processor of northern shrimp. One Maine processor, Claw Island Foods of Vinalhaven, guarantees "just like fresh" flavor and texture from its fully cooked, individually-quick-frozen (with their patented Sealock process), and ready-to-eat Maine lobster. "The frozen lobster will keep for one year in your freezer if not defrosted at any time," describes the company's brochure. Other Maine processors concentrate their efforts on packaging meat and freezing lobster tails. The processors typically supply cruise ships, resorts, casinos, supermarkets and restaurants. These companies have just formed their own industry group, Maine Lobster Processors.

The New Lobster Market

The lobster market has evolved to the point where it can now produce a year-round "quality product at affordable prices that competes well with other lobster substitutes [spiny, rock and European lobster, shrimp, crab, scallops and high-end fish, especially swordfish and salmon]," further explains Tourkakis. In the past when lobster was not available, many customers simply substituted another seafood for it. I'm told by knowledgeable people in the seafood industry that shrimp is the number one seafood in the world today.

Annual U.S. and Canadian landings, which have risen or stayed the same since 1990 to within the 170,000-million-pound range, have helped maintain the new lobster market. Maine has been the biggest U.S. producer lately, landing between 60 and 70 million pounds each year.

Today's Marketers

A slew of lobster marketers (including seasonal and year-round white tablecloth and eat-in-the-rough restaurants, fish markets, lobster pound operators, lobstermen co-ops, processors, traditional dock-side buyers, roadside shack and truck peddlers and even lobstermen) further divvy up the annual U.S. and Canadian lobster pie somewhere along the pathway between the catcher and the consumer. "It all comes out of the hatch," meaning the profits for handlers start with the lobsters being unloaded from the boats, and from here each handler tacks on their charges and earns their existences.

The sizes of these businesses can vary from small operations which might sell several thousand pounds of lobsters, to medium ones which handle about five million pounds a year like Mortillaro Lobster LLC and William Atwood Lobster Company, to giants like

East Coast Seafood, which wholesales over twenty million pounds per year. Like with lobstering, big is not necessarily better. What counts is the bottom line at the end of the year. Customers can also purchase lobsters on-line or over the telephone through many dealerships and have their orders specially packed in insulated cardboard boxes for land and air travel, promptly delivered right to their doorsteps.

The Yearly Picture Today

The lobster marketers generally want to quickly move their inventory, keep their cash flows and accounts going, their crews working, make room for more supply and hope for the best. "In this business, if you don't move the lobsters, you will lock your doors," says Anthony Burnham, long-time worker at Mortillaro Lobster LLC in Gloucester.

Although most lobster dealers know the yearly peaks and valleys in lobster marketing, they have also learned to expect the unexpected. Some larger dealers, processors and pound operators sometimes speculate on the market, and hold large volumes of lobsters, especially late in the season when prices can jump a dollar at a time. "I try to figure out where I will get the best return. Do I sell today, or do I pound, or do I freeze?" explains Spiros Tourkakis.

Typically, the yearly holidays create increases in lobster demand, which peaks at Christmas and New Year's, and during the summer when fine weather prompts outdoor activities. Prices usually rise around the yuletide season. The lobster market traditionally slows down right after New Year's, and prices briefly drop as many people diet right after the holidays and face their credit card bills.

Many marketers earn profits by reselling large volumes of lobsters at lower margins, while others do the opposite. This often-cutthroat business sees some of the players working together while others are rivals. Many former lobster suppliers from Canada and Maine have become direct competitors to their former U.S. middlemen customers; these suppliers now ship directly to their old customers' national and overseas markets. "One dealer's loss is another's gain" is very applicable for the lobster business.

"Mister Price sells the lobster," explains Vince Mortillaro. Being able to offer lobsters a nickel to a dime lower than competitors can mean the difference of a sale. Dealers like Mortillaro often spend much of their working day on the telephone trying to nail down deals.

I've always said the lobstermen have the best end of the deal, especially since we get paid at the end of the week and we usually don't have to worry about our lobsters once they have been dumped at the dealer's. But, the marketer's year, which involves lots of risks, also offers surprises and occasional opportunities to make big money. Some years are fishermen's years, while other years the dealers make out well.

The "boat price" for lobster can vary from port to port, often by twenty-five to fifty cents per pound at a given time. The boat price is usually highest around March, sometimes topping six to seven dollars per pound, when new-caught catches are at their lowest and their quality is at their best. Prices begin to drop after the Fourth of July, usually bottoming out in September and October, sometimes reaching the three-dollar per pound range. High catches of non-shippable lobsters are frequently prevalent during these months. Demand created by lobster processors often determines the boat price then. The price usually begins to ascend before Christmas, followed by a slight drop after New Year's Day. It begins to rise again by the end of January, and the pound keepers have their turn at making money. Sudden new-caught lobster supply shortages during holidays can spike the lobster price up fifty cents to a dollar per pound and earn dealers holding any lobsters huge profits. Successful veteran lobstermen have advised me over the years, "Fish for the lobsters, and not for the price." Experience has proved to me that day-by-day effort adds up at the end of the week, and something is always better than nothing at all.

The big players in the lobster market still seem to determine the lobster price so that they and other marketers can make something on the lobsters during each handling stage, while at the same time keeping the product moving and the markets alive. The boat price controls the consumer buying. Rising and falling lobster prices either open or close lobster availability to different income levels. The consumer ultimately controls the boat price. Not buying lobsters will ultimately force the price down to the point where it will spur buying again.

While some lobster ports pay catchers a straight crate-run price, others offer split boat prices for selects and hard- and soft-shell lobsters. Localized dealer "price wars" occasionally flare up, usually after one dealer has raised his price to lure more fishermen and his competitors follow suit, with the end result of the dealers losing and the fishermen winning. To avoid such wars, many dealers today pay the going boat price up front, but pay hidden bonuses at the end of the week to keep and attract fishermen. One Gloucester lobster dealer in 2004 tacked on additional payment for ten pounds for every one hundred pounds landed at the end of the week. Such dealer battles usually occur during supply shortages, or when a new dealer opens up shop. Like all other businesses, money talks in the lobstering business.

Lobstermen's suspicion over dealers and their boat prices still runs deep. A lot of fishermen believe some dealers still get together and wrongfully manipulate the boat price to their advantage. The late Ossie Beal, Beals Island lobsterman and president of the Maine Lobstermen's Association, is credited with starting Maine's first lobstermen's cooperative, the Beals-Jonesport Co-op, in 1971. Such cooperatives allow members to collectively market their catches for higher returns. In order to sell at these co-ops, members must first be accepted and then buy stock in the cooperative. Members must also agree to sell all of their lobster catches there. Co-ops also offer bait, gear, insurance, benefits—often at lower cost—and patronage refunds at the end of the year. Pigeon Cove Harbor lobstermen in Massachusetts started their own co-op, the Pigeon Cove Harbor Fishermen's Co-op, in the 1990s.

But in reality, the fishermen realize they need the lobster marketers. Without them, the lobsters wouldn't be worth anything, and there is no way most lobstermen can both harvest the lobsters and sell all of them.

Lobstermen have unsuccessfully gone on strike over low boat prices. The integrated lobster supply has rendered isolated strikes ineffective as affected dealers just pulled in lobsters from other locales to meet their needs. Other lobstermen, especially in Canada, have also tried to be dealers and crated up their catches during peak demand times in December of 2003 and 2004 to force the boat price up. This often backfires, resulting in high losses and poor-quality lobsters with chewed antennae and legs. Most dealers feel that the lobstermen should leave the marketing to them.

Lobstermen frequently hear their dealers crying, "We aren't making any money," yet the lobstermen see their dealers staying in business year after year, buying new equipment, especially trucks, and often expanding their operations.

The Marketing Risks

Despite the lobster market's safeguards, the state of domestic and foreign economies, the value of the U.S. dollar overseas, terrorist threats and attacks, airline strikes, the weather and even competition from other high-end seafood can quickly weaken the value of the lobster. "The headaches of lobster marketing are incredible," Tourkakis says.

The quality of the incoming lobsters can be one of these headaches. Lobsters might look nice from the outside and yet be half-full of meat. Other lobsters might be soft-shelled and weak and will not hold up, or their shell might be diseased. Strangely enough, up to a third of the catch of southwest Nova Scotian lobstermen were "softs" and of poor quality in November and December of 2003. Usually, their lobsters that time of year are of good quality. Many of these fishermen are now fishing further offshore with bigger, faster boats on the deep-water muddy bottoms where lobsters congregate and molt later in the year.

Ideally, dealers want lobsters that have some meat in them and are nice looking, lively and healthy. Dealers don't get paid for the dead lobsters. Tourkakis says, "You have to make sure the customers have a good experience with the lobster. Lobster quality starts on the boat," meaning that lobstermen need to handle the lobsters gently, quickly band them and keep them submerged in circulating seawater. A positive public perception of lobsters is important to keep the market going.

Getting paid can be a big problem for some of the dealers, too. "You sell on open terms," Tourkakis says. Dealers frequently have hundreds of thousands of dollars worth of lobsters out on the road. With many accounts, "You often have to wait thirty days before you get paid, plus the checks take another week to get here from Europe," explains Vince Mortillaro.

To guard against bad accounts, a lot of dealers purchase credit insurance and also the services of SeaFax, based in Portland, Maine. "For more than sixteen years we've been working on the credit side of the food industry. Over the course of those years, we've developed the most extensive database of financial and company-specific information in North America," states SeaFax. Besides offering collection service, SeaFax, via e-mail and fax, also gives "the most reliable up to the minute" company-specific credit information (including "bankruptcy data, bank and credit references, payment histories and credit risk indexes") in the industry.

Although most dealers have had long-term successful business relationships with many of their customers, some harbor painful stories of not getting paid. Bad European accounts helped do in the huge Atlantic Lobstermen's Co-op at Saugus, Massachusetts, in 1992. "I could retire on the hits that I've taken since I went into business with my father. This adds up to over a million dollars," says Mortillaro. Horror stories of the dealer getting stuck quickly reverberate within the marketing system.

The weather, which can affect both the lobster supply and dealer shipping, has created major headaches for lobster marketers. The October 31, 1990, No Name storm destroyed lobster gear and killed lobsters, crippling the supply of new-caughts along the east coast for several months and creating a demand that exceeded supply situation. As a result, there was a record eight dollar per pound boat price the following February. Since that 1990 Halloween storm, the lobster prices have only briefly reached seven dollars per pound, in April of 2003.

Another time, around Christmas in 1977, bad weather hit after East Coast Seafood "had packed and shipped out one hundred thousand pounds of lobsters on December 21. Flights were cancelled here and at other destinations. We tracked down some of the lobsters, some were returned and others were never found," remembers Tourkakis. Other flight cancellations have occurred because of bad weather or airline strikes, and dealers have had to pick up their shipments at the airport, unpack them, store the lobsters in their tanks and then go through the whole packing process again when conditions improve.

The millennium bust of December 31, 1999, and the 9/11 attack wreaked havoc on the lobster marketers. Many people, fearful of possible millennium-related computer glitches and terrorist attacks, stayed home rather than go out and celebrate. Months earlier, most dealers, believing the millennium a sure thing to make a lot of money on lobsters, began stockpiling live and frozen lobster inventory. The strong dealer demand for lobsters pushed the boat price up to around fifty cents per pound higher than usual. "We were sitting on hundreds of thousands of pounds and the price dropped two dollars per pound [after the millennium bust]. We had to remove the lobsters. This was a financial disaster," recalls Tourkakis.

The 9/11 Twin Tower terrorist attack temporarily killed much of the lobster market, already weakened by a poor economy, for about a week by both stopping airfreight and lobster demand, especially along the East Coast. Frightened Americans stayed home for months to come, crippling prime lobster outlets—restaurants, cruise ships, casinos. The William Atwood Lobster Company was the only dealer who actually had a lobster shipment, approximately 2,400 pounds, aboard one of the jets that crashed into the Twin Towers. Atwood was never reimbursed for this loss, since his insurance did not cover acts of terrorism.

Luckily, U.S. lobster catches were then off in most areas. Even so, a supply exceeding demand market existed, dropping the boat price by late November to the two-dollar levels, which had not been seen in ten years. Many lobstermen cut short their season. "Save the lobsters for next year when their price will be higher," they reasoned.

But by Christmas, the demand picked up and a few dealers who gambled, buying and sitting on large amounts of cheap Canadian lobsters late in the year, made big money when the pre-Christmas demand suddenly pushed the boat price up seventy-five cents per pound in less than a week.

"That's the lobster business; it's Las Vegas all over again. Times we win and times we lose," says Babe Smith.

VII
The Lobster Regulations

Close Reins

The United States and Canada keep close reins on their lobster fisheries under area management systems with many similar conservation and effort-control tools. According to the current scientific egg-per-recruit and growth-over-fished definitions, the lobster stocks in both countries are over-fished; yet much of Maine, Georges Bank and the southern Gulf and Scotia/Fundy continue to produce good lobster harvests. Regional lobstermen, who see firsthand what goes on out there on their lobster grounds every year, now work closely with regulators to craft conservation methods for their particular fishing areas. Besides the input of those fishermen, regulators also use the data from scientific lobster stock surveys and annual lobster landings to help them decide lobster regulations.

The regulations keep changing, complicated in the United States in the 1990s and early 2000s by right whale protection and inshore stock declines from the tip of Cape Cod south to Long Island Sound, and lately, the inshore section of the southernmost Gulf of Maine. The main U.S. lobster powers that be (the American Lobster Stock Assessment Peer Review Panel, the American Lobster Stock Assessment Committee and the Atlantic States Marine Fisheries Commission American Lobster Technical Committee) are now proposing "new median biological reference points as 'target and limits' for both fishing mortality and abundance to gauge the overall health of the resource," reported Janice M. Plante, associate editor of the *Commercial Fisheries News*, in the December 2005 issue. Canada has no national laws requiring the lobster stocks to be rebuilt within a said time as does the United States with the 1976 Magnuson-Stevens Conservation and Management Act as amended by the Sustainable Fisheries Act in 1996. Also, under the 1972 Marine Mammal Protection Act, it is mandated that whales, especially right whales, be protected.

The Lobster Regulations

The Old Days

How simple Cape Ann lobstering was in 1960. Most lobstermen thought of today and not tomorrow. The majority worked independently, lobstered from May through October, didn't belong to organizations and weren't technocrats. Many also took to the sea as an escape from shore-side responsibilities. There was no Marine Mammal Protection or Magnuson Stevens Fisheries Conservation and Management Acts. Seals, striped bass and pleasure boats were rarities. Economics, weather, lobstermen's unwritten laws and Massachusetts's general laws regulated that fishery. The grounds weren't peppered with traps, and much of the shoreline was undeveloped. All size lobsters were plentiful then. Lobsters brought the lobsterman fifty to fifty-five cents per pound, and cooked lobsters sold for around a dollar each at different eat-in-the-rough restaurants. There wasn't even the threat of offshore liquefied natural gas (LNG) terminals taking over fishing grounds.

In addition, anyone could get a lobster license by simply filling out an application and paying the five-dollar annual fee. Taking shorts or crickets (at that point, lobsters with carapace lengths under $3\ 3/16$ inches), eggers and female lobsters with V-notches, rarities most of the time then, were generally "no-no's" except for bringing home an occasional feed of shorts. Hauling, stealing, or destroying other lobstermen's gear, and fishing too early or too late in the day, were also frowned upon, then as well as now.

Although there were official game wardens, lobstermen were the biggest law enforcers. Offenders got either their boats or traps and buoys cut away, and even worse, boats sunk. Other unwritten lobsterman laws told you to fish within certain boundaries or lines from your coves or harbors or risk these same repercussions.

Although the wooden traps of that period were long-term conservation-friendly in that shipworms and storms would break down lost traps and eventually release their surviving contents, they weren't nice to lobsters in the short run. The traps, which lacked escape vents, often became too full of lobsters, which then crushed and killed one another. Plus, lobstermen handling them sometimes pulled out chains of shorts from traps, further damaging some.

Industry Goes Wild

By the late 1960s, the lobstering off Cape Ann had gone wild. Boat prices surpassed one dollar per pound; lobstermen were no longer looked at as second-class citizens; and everyone and his brother wanted to go lobstering either full- or part-time, especially teachers, policemen and firemen. Existing inshore lobstermen, aided by all those revolutions described earlier, often fished more traps during longer seasons, venturing further off the coast on bigger boats with larger crews. The lobsters were even getting pounded on their former tranquil mid-shore wintering grounds. The offshore trap and dragger fishery also blossomed along the continental slope, especially outside of Georges Bank.

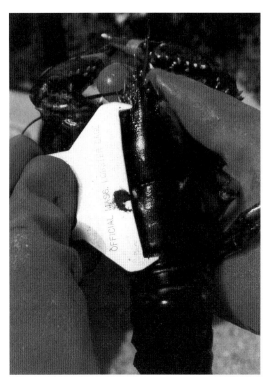

Here, a lobsterman measures a questionable-sized lobster with a gauge. Many fishing areas have both minimum and maximum sized gauges. Most just-legal-sized lobsters weigh just over a pound.

Despite Massachusetts beginning to limit the number of new entrants in the mid-1970s, annual catches doubled by the 1990s, often consisting of just-legal lobsters, estimated to be as high as 80 to 90 percent in some areas. The once-common larger brood stock females got scarcer inshore. The fishing pressure increased as the conservation plans were set in motion to bring back the striped bass, seals and depleted groundfish stocks, all main predators of lobsters.

Non-uniform lobster laws amongst many of the U.S. lobster-producing states and between the United States and Canada only complicated matters during this renaissance period. One area's loss was another's gain. A lobsterman in Massachusetts, which had a larger minimal lobster size than neighboring New Hampshire, could legally land Massachusetts shorts there. Also, Maine draggers and gillnetters, prohibited from landing lobsters in their state, could unload those very non-trap-caught lobsters in Massachusetts and Rhode Island. Today, Canadian lobstermen fishing the disputed Machias/Seal Island gray area right alongside U.S. lobstermen can take over-sized lobsters while their counterparts can't. Many of those lobsters are later shipped down to the United States. Maine has taken a stand and does not handle Canadian over-sized lobsters. This dispute has yet to be settled as of 2005.

New regulations were needed to insure this industry had a future. In the United States the Magnuson-Stevens Fishery Conservation and Management Act did just that, requiring "implementation of conservation and management measures to prevent over-

The Lobster Regulations

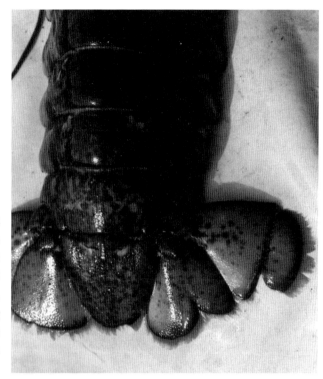

Most fishing areas also protect adult female lobsters by V-notching the second flipper in on the right-hand side of the tail. Both regulators and fishermen V-notch the females, especially when they are carrying external eggs. V-notched or punched lobsters are illegal to take in most fishing zones.

fishing." After assessing the three main stocks from Maine to North Carolina (Gulf of Maine, Georges Bank and Southern New England Outer Shelf (GBS) and South of Cape Cod to Long Island Sound (SCCLIS), the Atlantic States Marine Fisheries Commission (ASMFC) determined in 2000 that all areas were both "growth over-fished" and over-fished by its "egg-per-recruit" definition. The Commission defines growth over-fishing as when the "maximum yield in the fishery is not produced because of high fishing mortality on smaller lobsters." The Commission's egg-per-recruit over-fishing definition happens when "the lobster resource is harvested at a rate that results in egg production from the resource, on an egg-per-recruit basis, that is less than 10 percent of the level produced by an un-fished population." In other words, most of the females are getting caught up at legal size before they have a chance to add future lobsters to the fishery. For further clarification, the Commission defines egg-per-recruit as "expected egg production by a female in her lifetime. Usually expressed as a percentage of egg production in an un-fished stock." Maine's recent record catches, topped again in 2004 with approximately seventy-one million pounds of lobster, make one wonder if the Commission's assessments for all of the areas are accurate.

Since most lobsters are caught by traps in state waters, the ASMFC (and not the National Marine Fisheries Service, which regulates most fish and shellfish stocks in federal waters) got the lion's share of regulating lobsters, coming up with Amendment 3. ASMFC's "efforts are complimented in Federal waters by the National Marine Fisheries

The Maine Lobstermen's Association (MLA) has done so much over the years to protect both the lobster and the lobsterman. Past and present MLA officials (*left to right*) attorney Clayton Howard, past president Edward Blackmore and current president David Cousens enjoy a happy moment at the Maine Fishermen's Forum.

Service under the authority of the Atlantic Coastal Fisheries Cooperative Management Act of 1993."

Beginning in 1942, the ASMFC has served "as a regulatory body of the Atlantic Coastal states, coordinating the conservation and management of near-shore fishing resources." The states license their fishermen working the state waters, while the National Marine Fisheries Service licenses lobstermen working the federal waters or exclusive economic zone (EEZ). Today, local police and shellfish wardens, state environmental police and Federal National Marine Fisheries Service agents and the U.S. Coast Guard enforce the U.S. lobster laws. Atlantic Canada has its own Department of Fisheries and Oceans enforcement. Massachusetts fishermen have nicknamed the State Environmental Police the "Green Police," since they dress in forest green uniforms.

Amendment 3

The goal of Amendment 3, whose interstate rules are carried out by the separate states, "is to have a healthy American lobster resource and a management regime which provides for sustained harvest, maintains appropriate opportunities for participation, and provides

The Lobster Regulations

for cooperative development of conservation measures by all stake holders. The ongoing plan, when fully implemented, is designed to minimize the chance of population collapse due to recruitment failure." This goal is to be achieved by a series of steps by 2008.

Amendment 3 makes use of area management. The U.S. lobster grounds were split into seven lobster management areas (LMAs), each having a lobster conservation management team made up of area fishermen who make recommendations, later having to be approved by the ASMFC. My management area, Area 1, runs the coast from Maine to the tip of Cape Cod and also includes Stelwagen and Platts Banks and Jeffry's Ledge. Massachusetts has three LMAs—1, 2 and OC (Outer Cape)—and is adjacent to offshore Area 3—Georges Bank—and the outer Gulf of Maine.

According to ASMFC report number 29, each area is required to follow these specific

> *coast-wide requirements and prohibited actions which have been put into effect: Prohibition on possession of berried or scrubbed lobsters. Prohibition on possession of lobster meats, detached tails, claws, or other parts of lobsters. Prohibition on spearing lobsters. Prohibition on possession of v-notched female lobsters. Requirement for biodegradable "ghost" panel for traps. Minimum gauge size of 3 ¼ inches. Limits on landings by fishermen using gear or methods other than traps to one hundred lobsters per day or five hundred lobsters per trip for trips five days or longer. Requirements for permits and licensing. All lobster traps must contain at least one escape vent with a minimum of size 1 $^{15}/_{16}$ inches by 5 ¾ inches. Maximum trap size of 22,950 cubic inches in areas except Area 3, where traps may not exceed a volume of 30,100 cubic inches (roughly a trap no longer than 4 feet long). Traps now have to be tagged every year with a specific plastic-colored tag, also bearing the owner's license number and fishing area. Getting the traps ready for the upcoming season takes longer and longer with each passing year.*

Besides the standards, each area has the flexibility to come up with its own custom measures or "conservation equivalencies" to increase the egg production there, ranging from minimal gauge size increases, maximum gauge size decreases, zero tolerance of v-notch females, mandating lobstermen v-notching of all eggers and trap reductions. The overall plan has been modified five times through addendums, the last one in 2004. Addendum 4 addresses the lobster stock decline for Area 2 by immediately increasing the gauge size with more effort controls to take effect in 2005, while Addendum 5 limits trap strings to 2,200 and also increases the minimal gauge size to 3 $^{13}/_{32}$ inches for LMA 3. This action took affect in 2004.

Maine—Leader in Lobster Conservation

Maine has long led the United States and the world with effective lobster conservation and regulations, especially by protecting the brood stock lobsters. The state came up with the first double-gauge, having a 3 3/16-inch minimal carapace size and a 4 ¾-inch

Here, lobstermen attended a lobster-scoping meeting in Rhode Island. Most lobstermen today belong to an association or two with a common voice. Today's lobsterman often attends and speaks at meetings, especially those dealing with regulations that could affect his livelihood.

maximum carapace size in 1933. The switch to the current 5-inch maximum carapace length occurred in 1935. Lobstering has been the heartbeat for most of that state's coastal communities. The majority of Maine's lobstermen and powers that be, including its Department of Marine Resources, are committed to their lobster fishery. "We were conservationists before conservation was cool. We have always been the front-runner," says Jon Carter, a Bar Harbor, Maine, lobsterman. Maine has no Sunday lobstering from June through August. It does not allow any recreational fishing for lobsters, including scuba diving for them. The state has the only open and closed lobster season in the United States, which occurs at Monhegan Island. That open season runs from January 1 through June 30. Maine has long been the top lobster-producing state in the United States. This certainly proves they know what they are doing.

The 1,200-member-strong Maine Lobstermen's Association (MLA), which celebrated its fiftieth anniversary in 2004, deserves a lot of credit for maintaining Maine's successful lobster fishery. The association was started by Vinalhaven lobsterman Leslie Dyer, who presided there from 1954 until 1966, with attorney friend Alan Grossman, and later was presided over by Ossie Beal of Jonesport from 1967 to 1974, and Ed Blackmore of Stonington, who served as president from 1974 through 1991. Thomaston, Maine, lobsterman Dave Cousens took over as president then and is still serving in that position as of 2005. Pat White, a lobsterman from southern Maine, is serving as MLA's current CEO, along with Executive Director Patrice McCarron. All are dedicated officials. The state also has its own Maine Lobster Promotion Council.

Whale Protection

The Marine Mammal Protection Act of 1972, lately put into high gear by several conservation groups using the courts to force the National Marine Fisheries Service to come up with an Atlantic large whale take reduction plan (ALWTRP), has made lobstering more complicated. The ALWTRP is designed to reduce large whale entanglements and deaths resulting from commercial fishing activities. To do this, NMFS further carved up the U.S. lobster grounds into ALWTRP-regulated Lobster Waters, two of which are Northern Near-Shore Lobster Waters and Great South Channel Lobster Waters, and categorized the fixed and mobile gear fisheries from high risk to whales to low risk to whales.

Part of this plan requires fixed-gear fishermen to use one whale-safe gear option from a "lobster take reduction technology list," including breakaway buoys and the elimination of floating ground and buoy lines. The broad plan also includes special Dynamic Area Management (DAM) and Seasonal Area Management (SAM) programs. When three or more right whales are spotted in an area, "North of 40 degrees north latitude," fishermen are required to either temporarily remove their gear there, sometimes for up to a fifteen-day period, to make sure the whales are safely out of the area, or fish modified "low-risk" gear: trawls with no floating line, and marked only by one buoy with a weak link. The latter option applies to fixed fishing gear in the two seasonal area management locales "SAM west and SAM east" program. These have been popular right whale locations from early March through July. The gear moving option, which proved to be impractical and often dangerous, has become voluntary.

The whale regulations are one area where most lobstermen and regulators don't see eye to eye. Most lobstermen feel they are being unfairly targeted. Where I fish much of the year, northern near-shore lobster waters, I can see the bottom and have never had a whale entanglement in over forty years; yet, I'm required to fish whale-safe gear, which has so far cost me lots of unnecessarily lost buoys. A fellow lobsterman says, "A whale would ground out before it became entangled here." These whale rules will most likely only further intensify, since "Eight endangered right whale entanglement reports [were] received thus far in 2002."

"Ship strikes are the primary source of human-induced mortality to northern right whales. Five right whales were killed last year [2004]. Three were confirmed ship strikes, and the other two whales' cause of death couldn't be confirmed because of their badly decomposed states," Kate Sardi from the Whale Center of New England in Gloucester testified at a public forum in Gloucester on a proposed offshore LNG terminal which could be built about ten miles southeast of Gloucester.

Fishermen have borne the brunt of National Oceanic and Atmospheric Administration (NOAA)'s whale protection laws, the scorn of green groups and the TV and print media headlines, while it's been pretty much business as usual for the real culprits—especially military and shipping vessels. NOAA's whale protection version for them has been largely verbal.

NOAA broadcasts the whales' presence to these vessels over the National Weather Service's airwaves, along with urgings to proceed with caution and reminders that it is a federal offense to deliberately kill or injure whales.

Time is money for these ships, and even slow-moving ones can strike and grind up right whales. Fishermen have had to modify their gear with breakaway links, eliminate much of the floating lines and temporarily move their gear out of whale areas. A Boston newspaper headline read something to the effect in 2004, "Whale Fishing Gear Entanglements are on the Increase." At a time of declining fixed gear, when the remaining gear is whale-safe, many lobstermen read that as good news, meaning the increased entanglements mean growing numbers of whales. With more harvesters reporting whale entanglements and the presence of special whale-disentanglement teams, many fishermen also believe the entangled whales at least have a second chance.

The only sure way of protecting these whales is to shut the oceans down to everyone, including military and merchant ships, ferries and whale-watching vessels, which also contribute to the whale kills. You never seem to hear much about these. There is not a fisherman around who wants to hurt a whale.

Ironically, fixed gear (slime eel and lobster traps, gillnets and longline) and not mobile gear (trawls and dredges) is bottom friendly, while mobile gear is whale friendly and not bottom friendly. Green groups have formed campaigns on all sides of these issues. One can't help but wonder about their real motives. Fishermen can't win nowadays. Many fishermen believe that the big oil companies fund some of the conservation groups to push for such unfair and unrealistic legislation that is designed to force the fishermen out of business so as to eliminate future opposition to the energy companies setting up onshore and offshore oil and LNG terminals.

During April of 2005, Diane Borggaard and David Gouveia of the Protected Resources Division of NOAA Fisheries, Northeast Region, held a number of scoping meetings along the coast to get public input on six more large whale protection proposals designed to further decrease whale entanglements and death from fishing activities in the northeast. One of these proposals would ban all floating line by 2008. Borggaard, the large whale coordinator, and Gouveia, the marine mammal coordinator, stressed to the lobstermen in the audience at a Gloucester meeting, "We want to keep you guys working."

Canadian Lobstering

Through different branches, including Resource Management Regulations and Licensing and Conservation and Protection, Canada's Department of Fisheries and Oceans (DFO) controls the Maritimes' over-150-year-old lobster fishery. The DFO uses general (mainly from Atlantic Fisheries Regulations 1985) and area-specific regulations spelled out by the lobster license conditions for each lobster fishing area. Like in the United States, most of the fishing effort is done close to the land, but the inshore fishing effort has moved further off in recent years, often twenty to fifty miles offshore. Canada has long practiced area management for its lobster fisheries, beginning in 1899. The country also began to limit entry in 1968. According to a 1998 Overview of Lobster in the Maritimes conducted by the DFO in 1998, "There are over 6,400 licensed lobster fishers in the Maritimes. Over half of them are located in the southern Gulf of St. Lawrence. There are about two million lobster traps in the fishery; again, over half are found in the southern Gulf."

Canada's lobster fishery is split into four regions: Newfoundland, Quebec (northern Gulf of St. Lawrence), Gulf (southern Gulf of St. Lawrence) and Scotia/Fundy. Each region is further split into lobster fishing areas (LFAs): Newfoundland LFAs 3–14C, Quebec LFAs 15–22, Gulf LFAs 23–26B and Scotia/Fundy LFAs 27–41, which run from the northern tip of Cape Breton down the Atlantic and around into the Bay of Fundy to the United States border at Maine. Lobster fishing areas 1 and 2 are located in southern Labrador and are not shown on most charts. In 2002, the DFO established a new season for Area 38B in the Bay of Fundy, which includes the gray zone around Machias Seal Island, fished by both U.S. and Canadian lobstermen. This additional season runs from August 15 through October 31.

The *Profile of the Scotia Fundy Lobster Fishery* reports that "In 1998, the Minister of Fisheries and Oceans set egg/recruit target as a means of conserving and sustaining lobster stocks." In 1995, the Fisheries Resource Conservation Council (FRCC) "advised that there was a risk of recruitment failure in the lobster fishery unless measures were taken to increase egg production."

The DFO's regulations protect both its lobstermen and the resource with effective effort controls including limited entry, licensing and trap limits, trap-tagging requirements and open and closed seasons. In the Scotia/Fundy region Canadian lobster licenses fall under four categories: "A" (full-time), "B"(part-time), "P" (partnership) and "CC" (commercial communal). Each license category sets the maximum number of traps that can be fished in a LFA. Each license holder can only fish one LFA. Each LFA has a specific trap limit and opening and closing schedule. Most license holders fall under category "A."

The majority of the LFAs are closed from late July through October so as not to clash with peak fishing time for the U.S. lobster fishery. The Canadian system of open and closed seasons usually protects new-shellers, and the common late summer/early fall black-eggers, female lobsters carrying fresh eggs, while also enhancing the chance of mating and future egg production, since most females mate right after molting. The majority of the seasons last from eight weeks to approximately thirty weeks. Many LFAs have spring and late summer/early fall fishing seasons. Offshore Area 41, which includes German, Browns and the northeast part of Georges Banks, and Crowell and Georges Basins, is open year-round. This offshore fishery started in 1972. Many of these areas were formerly fished by U.S. offshore lobstermen until the 1984 Hague Line Decision put these waters under Canadian jurisdiction. There are no trap limits here, but only eight lobster licenses are issued for LFA 41. Nova Scotian lobstermen refer to this area as being "totally privatized" by one company, which owns all of the licenses there. This area functions under a total allowable catch (TAC), and each license holder has an individual transferable quota (ITQ) of ninety metric tons. Area 40, considered a critical lobster spawning area, is closed year-round.

Canadian regulations also protect the lobster resource in other ways. A range of minimum carapace lengths exist; most of Newfoundland's and Scotia/Fundy's LFAs use a 3.25-inch (82.50 millimeter) size—the same as in the United States, since much of the Canadian live lobster is exported to the United States. The gauge increase occurred in 1998. Many of the Gulf of St. Lawrence LFAs use a smaller minimum size, beginning at 2.66 inches (67.50 millimeters), since much of their catch stay in the country as canners, which are later processed for meat and frozen lobster products. This gauge size was raised in 2004 to 2.75 inches for Prince Edward Island lobstermen. Egg-bearing females cannot be taken in any of the LFAs, while the same is mainly true for V-notch females, except for LFAs 27 and 30. Although required by U.S. lobstermen working Area 1, lobstermen V-notching is voluntary in several of the Scotia/Fundy LFAs. In addition, except for two areas (LFA 30 and 31A), over-size females can be taken, and over-size males can be harvested throughout the Maritimes.

In addition, lobsters can only be harvested by traps and not by recreational fishermen. Traps cannot exceed a maximum size (approximately over four feet) and must be equipped with escape vents and ghost panels of some sort. Lobstermen can only land whole, live lobsters. Incidentally, the Maritimes have no right whale safe gear requirements, at least not yet. Most of the LFAs are closed and devoid of fixed gear when right whales inhabit these waters.

The rebuilding story of Newfoundland's Area 5 Eastport Peninsula lobster stock shows that the lobster regulations, along with a commitment from its fishers, do work. Area 5 recorded its lowest lobster landings in 1993. "We didn't wait for a disaster to strike," says area lobsterman George Feldman. The lobstermen there began V-notching females, while the minimal gauge size increased, and the grounds were closed to outside fishermen. "We did something right, and the landings have doubled since then," he says. No doubt similar actions will be taken to revitalize the Northumberland Strait fishery, whose landings have dropped 80 percent since the 1980s.

VIII
Lobster
The King of Cuisine

Chefs Jasper White and John Welch love cooking lobsters, and these masters also give their ideas as to why so many people find the American lobster scrumptious. "Lobster has that irresistible salty-sweet combination flavor; it's one of the great glamour foods," explains renowned New England seafood chef, restaurant owner and cookbook author Jasper White. His cookbooks, *Jasper White's Cooking from New England*, *Lobster at Home*, *Fifty Chowders* and *The Boston Fish Pier Cookbook*, offer traditional and creative lobster recipes.

John Welch, executive chef for the Point Sebago Resort at York Beach, Maine, further states, "The American lobster is the sweetest lobster in the world." Its meat is usually tender, moist, and easily chewable.

For some people, eating lobster can be not only a challenge full of surprises, but it can also bring back wonderful childhood memories of family get-togethers, cookouts and clambakes. Aided by crushers and picks, people like to dive into the whole cooked lobster and get all the meat out of its body. Many begin this process by first breaking off the walking legs and chomping on them to squeeze the meat out. They next work on the large claws, then the tail and lastly, the body, especially inside the carapace walls where there's meat inside the walking legs' sockets.

White's restaurants specialize in creative seafood dishes and New England fare. "We are a lobster restaurant. Everybody's favorite is pan-roasted lobster. That's the one I'm known for," says White, whose restaurants annually sell over one hundred thousand lobsters.

White's specialty dish can be either an entrée for two or an appetizer for four.

Pan Roasted Lobster with Chervil and Chives

One lobster, 1 ½-2 pounds
2 tablespoons peanut oil
¼ cup bourbon
¼ cup white wine
4 tablespoons unsalted butter
2 tablespoons either fresh chervil or parsley
2 tablespoons chives
salt and pepper, to taste

After placing the lobster, now split into four sections with its claws separated, into a 12-inch pan, the cooking passes through four stages: sauté it in peanut oil on the stove, then briefly broil it in the broiler. Then ignite the dish with bourbon and add wine, butter and herbs; season with salt and pepper. Finally, finish cooking on the stovetop. "The bourbon…adds a sweetness that mingles potently with the sweetness of the lobster. Fresh chervil imparts a hint of anise flavor to the lobster," says White. Either grilled shitake or Roman mushrooms or butter-fried vegetables are served with the dish.

John Welch of Bangor, Maine, who "grew up in the state of Maine and has been cooking lobster since age fourteen," won the Maine Lobster Promotion Council's Chef of the Year Award in 2001. Welch's prize recipe follows.

Native Maine Lobster on Herbed Brioche with Organic Greens, Fine Herbs, Shellfish Glaze and a Sherry Vinaigrette (serves 1)

One lobster

<u>Shellfish Glaze</u>
1 Maine lobster body
2 carrots
2 celery ribs
1 white onion
¼ cup brandy
6 cups vegetable stock
1 sprig thyme
1 pinch peppercorns
4 bay leaves

2 bunches tarragon
1 tablespoon whole unsalted butter
salt and pepper to taste

Herbed Brioche
1 tablespoon fresh chervil
1 tablespoon fresh dill
1 tablespoon fresh chives
1 tablespoon fresh tarragon
1 cup heavy cream
salt and pepper to taste

Vinaigrette and Salad
4 tablespoons sherry vinegar
1 shallot
1 teaspoon Dijon mustard
¾ cup canola oil
salt and pepper to taste
1 inch thick slice of brioche
1 cup organic mesclun mix
1 bunch chervil
1 bunch tarragon
1 bunch chives
1 bunch dill

Tarragon Oil
1 bunch tarragon
¼ cup canola oil

Remove tail and claw from the lobster body. Cook separately in boiling salted water. Cook tail for 3 minutes and claw for 4 minutes. Shell the lobster immediately. Use the reserved lobster body for shellfish glaze. In hot oil, brown a small dice of carrots, celery and white onion. Once the vegetables are caramelized, add lobster body and continue to brown. Once the lobster body and vegetables are thoroughly browned, deglaze the pan with brandy. Let the alcohol cook off approximately 2 minutes; cover with 6 cups of vegetable stock. Add a sachet of thyme, peppercorns, bay leaf and tarragon and let sauce simmer for 45 minutes; strain through chinois and reserve. Put sauce back on stove and reduce until the sauce has a glaze-like consistency (approximately reducing by ⅔ of original volume). Finish with 1 tablespoon whole unsalted butter; salt and pepper to taste. One lobster body should yield enough sauce for 8 servings.

For herbed brioche, use a standard brioche recipe. The key to the herbed brioche is an herbed egg wash using chervil, dill, chives and tarragon. Finely slice herbs using equal parts of each herb to yield ¼ cup. Steep the herbs by heating 1 cup of heavy cream and pouring over the herbs once the cream is hot. Blend in blender. While hot, add salt and pepper, shock in an ice bath and reserve for service.

For sherry vinaigrette, use the sherry vinegar; add shallot finely diced and Dijon mustard. Emulsify this mixture using ¾ cup of canola oil, salt and pepper to taste.

For assembling the dish, submerge the brioche in the egg wash and brown in sauté pan. Toss mesclun mix with sherry vinaigrette. Lay tail down on top of brioche, then mesclun mix, next the claw, followed by a nest of herbs including chervil, tarragon, chives and dill, seasoned with sherry vinaigrette. The plate will then be garnished with a shellfish glaze and tarragon oil. To make tarragon oil blanch and shock 1 bunch of tarragon. Blend in blender with chilled canola oil, strain through coffee filter and reserve.

Welch reflects on his thoughts on preparing lobster.

> *I generally cook the American lobster to showcase the lobster itself. Lobster is unique in the sense you can adapt all cooking methods such as boil, steam, grill, broil and even fry. Lobster can also be prepared as a cold salad or a hot entrée. It is important not to mask the flavor of the lobster in any dish. From a chef's standpoint, I believe that people who order lobster want to taste lobster; therefore, I generally keep the cooking process simple. Showcase the lobster, and you can't go wrong…I personally prefer to eat the traditional boiled lobster dinner with drawn butter, because this is what I have grown up enjoying.*

New lobster recipes and dishes constantly pop up, and the sky seems to be the limit with creative lobster dishes. There are many lobster dishes listed in cookbooks, brochures and menus besides the traditional boiled lobster, lobster salad and lobster roll.

The ambiance surrounding the lobster-eating experience unquestionably adds to its flavor and enjoyment. Jasper White's three hundred-seat Summer Shack restaurants in Cambridge, Massachusetts, and Uncasville, Connecticut (at Mohegan Sun), mimic the eat-in-the-rough picnic table-style of coastal lobster shacks.

Woodman's of Essex is another well-known lobster eat-in-the-rough restaurant that offers lobster menus in addition to fried and steamed clams. During summer afternoons,

Lobster, the King of Cuisine

Noted lobster chef and author Jasper White holds up two mainstays for his Summer Shack restaurants. His inland restaurants mimic traditional Maine lobster shacks where people can feast on lobster.

people often line up in front of the restaurant to wait their turns to order. This at least fourth generation old business, which does clambakes, too, is located in downtown Essex, Massachusetts, adjacent to the Essex River.

Tom Tedesco's Lobster Pool at Folly Cove in Rockport, Massachusetts, is a lobster shack along the rocky shore that abuts authentic lobster habitat. This restaurant is known for its lobster rolls filled with chunks of delicious meat. Here, the patrons can hear the sounds of gulls shrieking, lobster boat engines and waves lapping the shoreline, smell the salt air and see fiery sunsets. Tedesco's, White's and Woodman's restaurants are just three of thousands throughout the globe that serve the American lobster in its many forms.

Back in the early 1970s, as a guest for several trips aboard the former offshore Gloucester lobster dragger *Judith Lee Rose*, I experienced unique surroundings for eating lobster. During the winter and spring, this 117-foot-long trawler dragged the 1,000-foot depths of the Georges Bank outer canyons, including Corsair and Oceanographers, for lobsters. I had gone along for writing and photography material. For one January trip's Sunday dinner, the late cook Peter Manning, then seventy-six, cooked up a basket full of crushed and weak lobsters and prepared a number of lobster dishes from these including baked stuffed and boiled lobster, lobster rolls and salad, lobster stew and even fried lobster meat for the seven-man crew and this guest. The unique ambiance of eating the just-caught lobster over

People can dine on lobster just a few feet away from the ocean at Tom Tedesco's Lobster Pool at Folly Cove, which is known for its delicious lobster rolls. This seasonal restaurant also offers its customers beautiful views of Ipswich Bay and its spectacular sunsets.

two hundred miles offshore in the dragger's cozy fo'c'sle during January definitely added to the flavor.

But, like Welch, my favorite lobster dish is also the traditional boiled lobster dinner with drawn butter, with a special twist added: my best-tasting lobsters are those cooked right at the shoreline.

My family and I try to have around five lobster cookouts between Memorial Day and Labor Day each year at my childhood seashore picnic locale. From our house we trek the half-mile through the woods to the seashore, laden with eating and cooking utensils, including a boiling pot, cold drinks, lobsters, clams, butter, salad, cornbread and marshmallows.

Combing the seashore for clean driftwood, building a roaring fire and balancing the boiling pot filled with lobsters and a top layer of clams with about three inches of seawater at the same rocky crevice fireplace is part of the tradition. In no time, the boiling, frothing water will rattle the lid and spill over. The clams' opening signals the shellfish is done, usually requiring about ten minutes of intense cooking. Heat from the hot rocks melts the sticks of butter in a separate pot.

The steaming lobsters and clams are dumped on our makeshift table, a flat rock, where the rest of the food, eating utensils, paper plates and even small rocks used to crush the lobster shells are there for the taking. Everyone congregates around this rock for the feast. If you get hot, you can go into the ocean for a swim. The taste of this lobster meat just dipped in butter cannot be beat.

After toasting the marshmallows, the paper plates are burned before the fire is doused; the boiling pot is scrubbed at the water's edge with seawater and rockweed; and the shells are returned to the ocean. The trek home is much lighter. You also leave with a full belly, and another wonderful memory.

Over the years, I have learned these lobster preferences and dos and don'ts of cooking them. Friends who have gotten lobsters from me have often later remarked, "Those were the best-tasting lobsters I have ever eaten." Right off, some lobsters are meatier and sweeter than others. Chalk it up to experience; people in the business can tell a good lobster just by looking at it and picking it up. I prefer eating male lobsters. Their larger claws usually translate to more meat than females. Some people prefer the females because they also sometimes have lobster roe. Trial and error has shown me the new-shellers of late summer and early fall with semi-hard shells have the sweetest, most tender meat. You can tell a semi-hard-shell lobster by picking it up around the carapace and squeezing the carapace walls with your fingertips. The walls should be hard enough to require a little effort to squeeze them in. Also, you do not usually need crushers to break the large claws of these lobsters. The only drawback of these tasty lobsters is they have less meat than a rock-hard-shell one. The smaller lobsters, one pound to two pounds apiece, are generally tenderer than the bigger ones, generally three pounds and above.

Many people make the mistake of cooking lobsters in fresh water. Don't. The fresh water will get absorbed in the meat and dilute the flavor. If you don't have seawater, add plenty of salt to the fresh water so that it resembles seawater. Also, don't overcook the lobster, for this can make the meat mushy. I usually boil my lobsters for eight to fifteen minutes. You know the lobster meat has been properly cooked when it is firm, moist and tender.

Conclusion

What does the future hold for the American lobster and its industry, including the many thousands of lobstermen? The late Cape Ann lobsterman Eino "Yiksi" Leino probably had the best answer: "Time will tell." During the start of my lobstering in 1960, I would never have dreamed about what has unfolded in the lobster fishery in the forty-six years since then.

The hunted and the hunters have held their own for the most part as of 2006, but nothing is ever etched in stone. Man-made and natural forces continue to threaten area lobster stocks and eventually those harvesters, as they have in U.S. lobster fishing Areas 2 and 6 (inshore from Cape Cod south to Long Island Sound) and Canadian lobster fishing Area 25 (Northumberland Strait) where catches have plummeted over 50 percent. The lobsterman's way of life is also threatened by hidden forces that have nothing to do with lobster abundance. If the lobstermen go out of a certain area, so do their supporting shore-side infrastructure, their way of life in the community, along with the trickle-down effects of their dollars into the local and state economies.

Right off, rising energy prices, which began in earnest in late 2004 and are continuing in 2006 with no end in sight, have already seen diesel fuel and gasoline surpass the two-dollar per gallon mark. These increases could threaten the economic feasibilities of many lobstermen's operations. The escalating prices eventually get passed on to just about every aspect of the lobster business, including the wire and netting in traps, rope, buoys, paint and bait. Despite the rising operating costs, lobster prices frequently do not keep pace with these and often actually go down when the supply increases. When the lobster price is too high, many people just won't buy, and the only way the dealers can move this perishable is to lower the price. Fortunately, my lobster business is very economical to run; I burn only about seven gallons of gas a day. Many of the lobster boats with medium horsepowered diesels burn five to six gallons of fuel or more an hour during their operations. But the big engine boats with five hundred horsepower or greater diesels frequently use seventy-five to one

Conclusion

hundred gallons of fuel during a twelve-hour day. Either way, the gas and fuel costs add up quickly.

Three experts in the lobster fishery (Bruce T. Estrella, senior marine fisheries biologist/project leader of the Coastal Lobster Investigations Project, Massachusetts Division of Marine Fisheries; Carl Wilson, who has already been described and made some earlier comments; and veteran Long Island Sound lobsterman Nick Crismale from Guilford, Connecticut) give their thoughts on the biggest threats to the American lobster today. Wilson's comments only apply to Area 1, with which he is most familiar.

Estrella explains,

> *The greatest threat to the American lobster is man. Some may consider this threat in the form of man's impact on the environment through coastal development, pollution, disruption to or occlusion of critical habitat by construction-altered sediments and siltation. However, despite unfavorable environments, predation, coastal storms, disease and other acts of God which can debit populations of marine life in the form of what we* [fishery scientists] *refer to as natural mortality, man's commercial fishery for this clawed delicacy represents the largest source of removal of lobsters from the sea…This should not imply that man can be held above predatory status since a consciousness about maintaining this species would prevail; shouldn't it? Instead, resistance to sound resource management philosophy including the need to control and reduce excessive removal rates which jeopardize adequate egg production, sustenance of abundance, improved size structure and optimal meat yield has characterized most of the inshore industry's response to attempts at meaningful American lobster fishery management. It is difficult to find man's wisdom in maintenance of the instability associated with a recruitment-dependent fishery with untethered expansion of fishing effort which sustains vulnerability of each recruiting year class to the elements in the absence of a buffer against any recruitment shortfall.*

Wilson believes,

> *Clearly the fishing industry is placing a heavy burden of exploitation on the lobster resource. The length structure has probably been reduced in the inshore waters since at least the mid-1930s (as far back as I can find length data in Maine). Interestingly, there is very little evidence that the fishery has influenced the abundance of lobster positively or negatively over the years. In the last twenty years the industry has adapted to an increase in abundance; if we were to return to "normal" levels through a natural decline (poor recruitment, disease, increased predation, etc.) there would be economic chaos, but the stock would probably be okay, if effort could be reduced back to what it was in previous years. The ability to measure and reduce fishing effort is very tricky with lobsters, as the relationship between traps and removal has not been clearly demonstrated, so if there is a stock change we may not be able to react to it through management. When put in terms of "the greatest threat," the fishing industry can be fingered, largely because that is what we*

can control, history has shown that industry probably has less to do with the population than we may know. I do not consider natural swings in the population a threat to the lobster resource, but a threat to the lobster fishery.

This letter to the editor by Nick Crismale, published in 2003 by both *Commercial Fisheries News* and *National Fisherman*, expresses his fears for the lobster fishery. Crismale has experienced first-hand a lobster and lobsterman catastrophe in Long Island Sound.

The Northeast lobster industry should wake up and realize the catastrophic threshold that its resource is facing. I have been a lobsterman in the Long Island Sound for thirty years and have experienced the "ups and downs" in catches just as all other lobstermen in all areas have felt, but the recent declines in lobster populations compounded by the additional problems of shell disease would lead a prudent person to question whether there is more at work here than over fishing or Mother Nature.

At a recent workshop in Connecticut, our Department of Environmental Protection (DEP) met with lobstermen to discuss potential programs for the recovery of the lobster fishery in Long Island Sound. A wide range of proposals to sustain the fishery were discussed, including moratoriums, area closures, and gauge and escape vent increases. The bottom line is that no matter what efforts are made it is still uncertain whether our lobster population can be sustained in this environment. A recent DEP study determined that there is an approximate rate of 35 percent natural mortality in Long Island Sound. What happened since the summer of 1999 to increase the threat of extinction of our fishery? Warming temperatures? Hypoxia? Or maybe toxic poisoning from pesticides. All of these theories are being studied and, hopefully, by January 2004 the results of three years of this research will be available.

My concern is whether this research will yield enough results and evidence for our environmental "managers" to make the necessary decisions to bring back our fishery. Six and half million dollars have been spent on research. As an industry representative from Connecticut sitting on the steering committee for these various proposed projects, I am very doubtful that any findings or conclusions by the researchers will be enough to prompt our managers to develop a plan to save our fishery.

I recall talking to industry representatives from Rhode Island, Massachusetts, and Maine as well as Canada, about the devastating impacts from pesticides such as malathion, resmethrin, sumethrin, methoprene, and others used for mosquito and moth control. I could see in their faces they thought that Long Island Sound has an area specific problem, and they never believed that it could ever affect the lobster population in their areas.

Well, something is responsible for the declining lobster population and increased presence of shell disease. With the heightened media scare of West Nile virus and the local and regional health departments' use of pesticides to control mosquitoes, the lobster industry should be cognizant of its ability to become collateral damage as part of this war on the mosquito. No one knows how these chemicals interact with each other or how, in some cases, the metabolites from these chemicals impact our environment. Metabolites can be even stronger than the original chemical applied.

Research should have some focus on the toxic effects of these chemicals. Lobstermen should make themselves aware of the pesticide programs in their areas and the chemicals being used. These chemicals have more than the ability to kill; they can affect life cycles as well as reproduction. It is important to remember that lobsters are from the same family as insects being targeted for annihilation.

One last thought on research—it is ironic that a man named Paul Muller was hailed as a savior when he discovered DDT and, in 1948, was even awarded a Nobel Prize. But like CFCs, DDT was at the same time invisibly attacking the foundation of life.

Makes you wonder about research. Researchers are always looking to solve the immediate problem, only to be blindsided by something they never anticipated.

I largely concur with Estrella and Wilson that the fishing industry is the biggest taker of lobsters, but with one exception. Those lobsters are size-specific—each weighing at least over a pound, and for U.S. fishing Area 1 and Maryland's state waters, not weighing much over four pounds—and for all areas, are females free of external eggs and V-notches. The industry and regulators make sure the rest of the trapped and netted lobsters are returned to the ocean to grow and to repopulate the stocks.

For many U.S. lobster stocks today, the fishing pressure is an annual constant created by more lobstermen fishing year-round, with traps that can hold more legal lobsters for long periods. Scientists and regulators fear lobstermen are so good at their jobs today that they are removing the majority of the new recruits every year. Fortunately for the lobster stocks, good conservation rules are in place, and not all lobsters feed at once, and not all of those that enter a trap get caught. Most stocks have kept up to the heavy harvesting constant, at least until lately. Winter and early spring hibernation, and aestivation, offer many lobsters a break from the fishing pressure in the United States. Of course, Canada's open and closed seasons do the same. But John V. O'Shea, ASMFC executive director, recently expressed these dire concerns of the American Lobster Stock Assessment Peer Review Panel about the U.S. lobster stock.

It is the future that leads to the greatest peril for the lobster resource. It would only take a sequence of two to three years of poor recruitment to collapse any component of the lobster resource, and the appearance of extremely low recruitment in recent times in some areas is a cause of concern if not alarm. Until the harvest strategy is revised to provide a buffer of mature adult spawners to cover the bad times, lobster fishery management is a time bomb waiting to explode, its fuse lit by recruitment failure.

James M. Knott Sr., along with many U.S. lobstermen, believes the year-round trap fishing actually helps create more lobsters by providing year-round food, which, in turn, causes the lobsters to grow and multiply more. But not all lobstermen, especially the

Canadian counterparts, give credence to Knott's theory since their lobster fishery is mainly based on open and closed seasons, and many of their areas have had recent record catches like those of the U.S.

Extremes of weather such as turbulence or heavy rains and lower salinities, or warm water and low oxygen levels, will continue to stress, or worse, kill, a percentage of an area's stocks, especially the young, which are generally in close to shore and can't move as quickly.

I feel that most stocks can handle the regulated fishing pressure and the temporary meteorological extremes. Longtime experience has repeatedly shown me that lobsters have legs and tails, and that many can actually tell when bad weather is coming long before we can, and they vacate unfavorable conditions and return when normalcy is back. But not all are that lucky, especially when storms hit by surprise or intensify and turn out to be much worse than anticipated. The Blizzard of '78 and the October 1991 No Name storm washed ashore thousands of dead and dying lobsters along many of Cape Ann's beaches.

Offshore winds in the summer and fall can blow lobster larvae seaward to unfavorable bottoms lacking hiding places like rocky crevices, and greatly diminish their survival chances. But these winds can also transport babies from their hatching areas to other favorable locales along the coast, giving those areas a boost in the future.

Shell disease is one of the industry's biggest threats and unknowns. It appears to be getting worse in many inshore areas from Cape Cod south, sometimes making up 20 to 30 percent of the lobster catches there. Shell disease was first reported in 1996 at Long Island Sound. It was also found in the lobsters living in the offshore canyons off New York in 2000.

Although scientists generally agree that a host of microbes, mainly bacteria, cause the shell disease by breaking down and consuming the shell components, they still don't know for sure what weakens the lobsters in the first place and makes them susceptible to shell disease. Man-made pollutants could possibly be to blame.

Dr. Bob Bayer, of the Lobster Institute at the University of Maine in Orono, believes "that shell disease for lobsters with a good immune system and good nutrition is not a problem." The disease doesn't kill the lobsters directly, but it further weakens the lobsters' immune systems and makes them prone to other life-threatening ailments and stresses. This disease is common in egg-bearing females. There are two types of shell disease, the rosette and the isolated pits. The rosette type looks as though someone had sandpapered an area to the point where only a rubbery skin was left to the shell. The isolated pits resemble corroded metal.

Many scientists believe that man-made chemical compound stresses, especially alkyl phenols, which are widely used in the U.S. to make alkyl phenol ethoxylate detergents (especially liquid laundry detergents), could be causing the shell disease in many of the stricken areas. These alkyl phenols end up in the lobster's blood, and from there they probably weaken the lobster's immune system and pave the way for shell disease. The alkyl phenols enter the marine ecosystem through runoff, but more so probably from sewage treatment plant effluents that exit from offshore outfalls.

Over the years, I have seen shell disease come and go inshore off Cape Ann, at least prior to 2004. In 2003 my traps often came up with two to three infected lobsters every haul. Ironically, I saw the fewest number of shell disease lobsters in 2004. The biggest incidence occurred briefly in December that year when a body of pound-and-a half- to two-pound males moved in and took up residence along the Rockport breakwater. I might have trapped a dozen or so then. This led me to believe that the shell disease lobsters also travel as a group.

Lobsters are further threatened by more human activities on the land, as we attempt to chemically rid lawns, gardens and even neighborhoods of insect and weed pests. The petrochemical industry offers quick chemical fixes that are usually either sprayed or powdered on. Most of the insecticide companies say these chemicals are species-specific and will break down in the environment. The U.S. Government even carries out widespread pest eradication programs, especially for mosquitoes and the brown-tailed moth. These chemicals are frequently used when lobsters are the closest in to shore, molting and mating, and the young are hatching and growing there.

Gone are the days of manually removing the unwanted and using commonsense solutions to seasonal problems. What's in the backyard has a way of sinking in the soil, being picked up by moving water and finding its way to the ocean and lobster nursery grounds. I also am aware that lobsters are arthropods, just like the insect pests, and I doubt that many poisons are species-specific.

The population also wants a steady supply of energy to fuel vehicles and boats and to light up, heat and cool businesses and homes. Meeting those energy demands involves transporting lots of oil and liquefied natural gas atop the ocean in tankers and barges and below it through trans-country pipelines, all translating to risks of potentially deadly oil spills and lobster habitat alterations. Despite industry assurances and added safety measures brought about from past mistakes, human errors and mechanical malfunctions will result in future oil spills, and who knows what other kinds of energy-related disasters.

Two significant oil spills off the northeast coast have already happened. One spill involving approximately ninety thousand gallons of No. 6 fuel oil occurred in April 2003 when the 376-foot barge *B120*, owned by Bouchard Transportation Company, hit a submerged rock in Buzzards Bay while being towed by a tug. Fortunately, most of this oil did not sink to the bottom, and it was quickly cleaned up with minimal damage to the environment.

The second oil mishap killed approximately nine million lobsters along Rhode Island's south shore in January of 1996 after the 340-foot-long barge *North Cape* washed ashore during a storm and leaked out an estimated 828,000 gallons of No. 2 home heating oil from its 3.9 million gallon load. The March 1996 issue of *Commercial Fisheries News* reported,

As of mid-February, the full extent of the natural resource damage was still being assessed. However, reports of beaches covered with literally tons of sea clams and tens of thousands of lobsters…pointed to an environmental disaster. "It was unbelievable," said one biologist who spent days walking the shore, counting and cataloguing the dead animals. "It was like the ocean threw up on the beach and then did it again and again." As of February 15, some 256 square miles of ocean off Rhode Island's southern shore, including the waters around Block Island, were closed to all shellfish, finfish, and lobster harvesting, and another 40 square miles south of Block Island were in the process of being closed, according to David Borden, DEM assistant director for natural resources protection and development.

The fishing grounds, part of Area 2, have since been reopened, but the lobstering has been nowhere near up to par. Many of the lobsters in this area suffer from shell disease. A lobster-seeding project, which has involved the gradual release of over one million egg-bearing females, has been conducted.

Despite being offered $8 million annually from several energy companies to build an over $300 million liquefied natural gas terminal in Harpswell, Maine, the town voted to reject their offer in 2004, fearing this would hurt their lobster fishery. The project would have laid down miles of submarine piping on productive lobster habitat.

Abutters and users of Massachusetts Bay are now finding themselves in a similar situation in 2005. Excelerate Energy LLC and Tractebel LNG North America began proposing to build offshore LNG terminals about ten miles southeast of Gloucester right in the middle of prime lobster and groundfish grounds designated as Block 125. It's adjacent to the Stellwagen Bank Marine Sanctuary, the Boston shipping lanes and the Massachusetts Bay disposal site.

The project would not only displace fishermen, pose safety risks to population centers, threaten endangered whales which seasonally occupy the area and disrupt the bottom with pipeline needed to transport the gas to the shore facility, but it could also possibly change the ecology of the area. The off-loading tankers discharge huge amounts of hot water during the liquid gas vaporization process needed to pipe it ashore, and this could alter the water temperature at the site. The over nine hundred-foot-long tanker hulls could foul the area with copper, too.

"I've been fishing there off and on for thirty-five years. Block 125 is one of our prime grounds. Fishermen have fished there for over four hundred years. We have been sacrificing and nurturing the ocean back to life the past twenty-five years, and now these guys come along and say they want that piece of bottom," says Gloucester vessel owner-operator Ignazzio Sanfilippo.

But the biggest manmade lobster disaster so far, at least in the minds of approximately 1,200 Long Island Sound lobstermen, occurred here in September 1999 when widespread mosquito spraying took place in New York and Connecticut to ward off the threat of West Nile virus. The spraying was followed by Hurricane Floyd, which dropped torrential

Conclusion

rains, resulting in massive runoff, including the insecticides, into the largely land-locked Long Island Sound. Before the storm, lobstermen hauled traps aboard with live and flipping lobsters; the same traps came up with dead and dying lobsters after the storm.

After four annual Long Island Sound lobster health symposia in Stony Brook, New York, scientists concluded a synergism of causes was responsible for the lobster die-off. An article in the November 2004 issue of *Commercial Fisheries News* by associate editor Janice M. Plante explains the scientists' findings. "They found that the die-off was caused by a combination of unfavorable circumstances, including warm water temperatures, low dissolved oxygen, ammonia and sulfide toxicity, storm action, rapid turnover of the sediment-water interface, heavy rainfall, and extensive freshwater runoff."

Jack Pearce, a retired NMFS biologist, offers this explanation.

> *Substantial evidence exists that an unusual synergism of factors pushed the western Long Island Sound lobster population far out of equilibrium with its environment…*
>
> *The lobsters were subjected to sustained and increasingly hostile conditions. Their immune systems were overwhelmed and unable to compensate, and many died as a result. The lobsters in western Long Island Sound would have suffered extensive mortality in 1999 in the absence of any application of pesticides.*

Most of the Long Island Sound lobstermen don't buy any of the scientists' explanations. "Low oxygen levels and high water temperatures have always been a problem in Long Island Sound, but have never caused die-offs like the one seen in 1999 and 2000. The one thing that is different is the pesticides. If they hadn't sprayed like they did, guys would still be fishing," says lobsterman Nick Crismale, who has fished Long Island Sound for thirty-five years.

Three of the area lobstermen, including Crismale, filed a federal class action suit in 2000 against the five pesticide companies that supplied the mosquito spray. Two of the companies worked out a settlement in 2004. Three cases are still pending. The lobstering in the western end of Long Island Sound is still poor. Many fishermen have since either sold their businesses or everything, including their homes, and relocated to other states. Today, up to 20 to 30 percent of Long Island Sound's lobsters are also afflicted with shell disease.

Big non-chemical threats to the lobster industry today, especially in the United States, ironically come from the Marine Mammal Protection, the Magnuson-Stevens Conservation and Fishery Management and Conservation and the Atlantic Striped Bass Conservation Acts. Besides protecting and restoring the creatures they were designed for, these national and inter-state acts have also brought back most of the lobster's top natural predators, especially seals and striped bass. The catches of U.S. southern lobster fishing areas 1 and 2 are dropping, probably because of this predation. The best twenty recent years of lobstering, which Carl Wilson also referred to, occurred when these predators' numbers were low.

Seals and bass were rarities off Cape Ann until the 1980s. The seals now occupy those waters en masse from October through April, while the bass arrive in late May and usually stay until October.

Many inshore lobstermen from New Hampshire southward believe the striped bass is the biggest threat to lobsters today. The lobster's once-safe shallows, where they molt and mate and the babies hide from early summer to early fall, are no longer safe, thanks to the bass. These voracious predators will swim in water so shallow after prey, that their bellies will ground out on the bottom, and their dorsal fins will break the surface water. The inshore lobsters, which sometimes molt in the open under seaweed or in holes on harbor bottoms, are sitting ducks to the bass. The bass typically suck in and swallow the lobsters whole.

"We have a gauge and escape vents. The predators don't. It's the size of their mouth which regulates what size lobster they will eat," says Pigeon Cove lobsterman Robert "Mo" Morris. Luckily, for more northern inshore lobster stocks, the colder water has largely kept the stripers out of those areas.

Seals and bass often wait under working lobster boats for easy meals of tossed-back lobsters. This is why lobstermen strongly believe today's traps should have good escape vents to easily let the little ones escape and keep them on the bottom.

Seals are a growing threat, too, especially to the more-northern lobstermen and their lobster stocks. Besides eating lobsters, the seals also spring open the traps' doors and rob the traps of bait and probably lobsters. Lobstermen who fish traps ten miles offshore over three hundred feet down have even told of having their traps sprung open and bait stolen by the seals. The seal's sense of smell must be terrific.

I witnessed a seal surface near my lobster boat one January day in 1999 right in back of Pigeon Cove Harbor with a lobster right in its mouth. I was hauling traps when a loud "crunch" caught my attention. I turned around, and there was a seal bobbing vertically with a half-swallowed lobster, claws sticking up, in its mouth.

The naïve often say, "The seals are so cute!" That cuteness soon wears away when you have to deal with them regularly and they are taking money out of your pocket. People used to say the same thing about deer. Most fishermen look at seals the same way that many people view a rat in their house or a mosquito on their arm.

Towns and cities used to have bounties on seals for a reason. They paid hunters five dollars per seal nose. The old-timers knew what they were doing shooting the seals. The day will come when U.S. regulators will have to deal with the exploding numbers of seals, which now lack few natural predators other than the large carnivorous sharks and killer whales. Seals also steal fish out of gillnets, tub trawls and fish traps. No doubt, they are threatening the recovery of some groundfish stocks, like cod, too. Canada has lately realized this, and fortunately, that country now allows a limited harp seal hunt. That seal's population is estimated to be greater than five million in Canada.

Many of today's lobstermen are also threatened from the land. Beginning in the 1970s, the shoreline, waterfront and ocean have been discovered big time. This trend is spreading further north where property values are still reasonable. Out of town money and leisure time now abound. The coastal demands for living, working and mooring spaces have exceeded their supplies. Much of the coastal land, housing, business and mooring supplies are tied up in the lobster industry. The pressure, especially from the real estate and recreational boat/marina industries, is on for owners of waterfront property, including homes, fishing shacks, open space, docks and even whole coves, to sell for the big bucks. Selling for the big bucks is like the person who won a $20,000 lottery and ended up pocketing $14,000, while Uncle Sam took the rest. Once any property has been taken out of the lobster industry, it's basically gone forever. Few lobstermen can afford to get back in.

Fortunately, many towns and cities have marine industrial use zoning for their waterfronts along with government-owned piers, docks and marinas that favor commercial fishermen, including lobstermen. Here, lobstermen can park their vehicles, dock their boats, load and unload catches, supplies and gear. Without these shore-side facilities, many lobstermen couldn't operate. Most towns and cities now have long waiting lists for moorings.

Big outside bucks will gradually change the whole character of a fishing village as new owners and investors buy up space, soon razing or rebuilding the existing structures into mansions. Quickly, the "No Trespassing" signs and fences will go up, and complaints over noise and foul smells will start. The demands for municipal services, including new sewer systems, followed by escalating property taxes to pay for these and rising real estate values will occur. In no time, the area will get too expensive for many of the local people to remain, and especially for their offspring to settle there. One mid-coast Maine fisherman told me he paid $70,000 for a new home in 1978. The site had a distant water view. He was offered $500,000 for that same property in 2003.

Many of the above scenarios are happening right now on Cape Ann, especially after Massachusetts forced Gloucester to install sewers to stop coastal pollution so that more clams could be dug on area clam-flats. From what I've seen, this expensive task has done more harm than good. My house in Lanesville received a sewer connection around 1995. Although this sewer project has eliminated most of the raw sewage from entering brooks and streams and the coast, it has dried up ground water and opened the door for more development, noise, traffic and insecticide and fertilizer pollution. House values have sky-rocketed, more mansions are being built on the waterfront, taxes have risen along with other municipal bills, the roads have become more pothole-filled and the city's infrastructure has deteriorated. With the increased traffic, it seems as though there's always somebody on your tail nowadays. The recreational boating and real estate industries are now lying in wait for the commercial fishing industry and its shore-side infrastructure to collapse.

After primary treatment, which includes the addition of chlorine, all of Gloucester's treated sewage is later piped and dispersed off Gloucester Harbor into Massachusetts Bay. This certainly isn't nature's way. I'm sure all of this fresh water and chlorine can't be helping the ecosystem there.

I've already fought and helped win two battles on Cape Ann to preserve the working waterfronts of Lane's Cove and Pigeon Cove Harbor, and that includes public access to them, too. The lobstermen of both of these once-privately-owned granite block harbors on Cape Ann faced displacement by new private interests.

Around 1964, Boston University tentatively selected the privately-owned Lane's Cove as a site for a new marine lab and attempted to buy the property from its owner, the Lane's Cove Pier Company. This facility would occupy the area of the cove where most of the approximately twenty lobstermen, including myself, moored their boats. I was a high school sophomore at the time. Fortunately, the cove had been used as a public landing, marked by a stone with a copper seal, since colonial times, also verified by the Massachusetts Land Court. Soon a public campaign to save Lane's Cove got underway, including rallies, speeches, petitions, special hearings and the formation of the Committee for the Preservation of Lane's Cove Landing Place, which used the cove's public landing as a legal weapon to prevent Boston University from taking over Lane's Cove as a marine lab site.

B.U. stopped its drive on October 16, 1965. Helped by the state and federal government, the City of Gloucester's Conservation Commission purchased the approximately seven-acre property, including two piers, two breakwaters and the beaches and flat rocks between them for $50,000. The cove is now home to about twenty lobstermen and thirty recreational boaters. It's also widely used by the public, including artists during the summer.

An article of mine, which appeared in the August 2002 issue of *Commercial Fisheries News*, best describes the second event—the Battle for Pigeon Cove Harbor.

The Battle for Pigeon Cove Harbor started in 1993 when a private marina blocked land access to most of the cove's 60 mooring holders, many of whom were commercial fishermen. The cove had long been the anchorage site for Rockport-based small draggers, tuna and lobster boats, and pleasure craft. The town had maintained access through a lease arrangement with the private owner of the 9.6-acre Pigeon Cove property, which included the former Cape Ann Tool Company.

But in the late '80s, ownership of the land was transferred to heirs who formed the Pigeon Cove Land Corporation (PCLC). Despite negotiations over several years, the town was unable to work out a purchase agreement with PCLC. Instead, the owners leased the property to a private marina, which intended to build an upscale facility at the site.

The marina group, in what was perceived as an effort to drive existing boats out of the cove, put exorbitant fees on use of the shore land access to the moorings. Refusing to pay the fees or step aside, the cove's mooring holders formed the Pigeon Cove Boatowners Association (PCBA), led by officers Presidents Albert Johnson and Al Olson, Vice Presidents James Waddell and John Halmen, Secretary Diane Nelson and Treasurer Fred Nelson. The group threw a monkey wrench into the marina's plans by not moving and continuing their fishing businesses by shuttling to their boats in Pigeon Cove Harbor, a federal anchorage, from adjacent coves.

Conclusion

The PCBA hired attorney Tom Delaney of Delaney, Scuderi, and Goddard of Salem and sought the help of local, state, and federal officials. The group held fund-raisers, media campaigns, and protests (all peaceful), making their plight well known and a political liability to the property owner, PCLC. The men also uncovered illegal asbestos handling by the marina, which ultimately led to its demise.

Finally, in August 1994, the PCLC agreed to sell the property to the local boat owners. PCBA bought the approximately four-and-a-half-acre working wharf with upper and outer breakwaters for $300,000. The PCBA has since transferred ownership of the property to the Town of Rockport, Massachusetts, on June 26, 2002, ensuring that the cove will be public forever. A $150,000 grant from the Massachusetts Seaport Bond Bill Council helped pay off the mortgage.

Time never sits still, and you can never take anything for granted.

Bibliography

Bliss, Dorothy E. *Shrimps, Lobsters and Crabs: Their Fascinating Life Story*. New York: Columbia University Press, 1982.

Bowers, George M. *Bulletin of the United States Fish Commission, Vol. XIX for 1899*. Washington, D.C.: Government Printing Office, 1901.

Collette, Bruce B. and Grace Klein-MacPhee, eds. *Bigelow and Schroeder's Fishes of the Gulf of Maine*, 3rd ed. Washington, D.C.: Smithsonian Institution Press, 2002.

Commercial Fisheries News. Stonington, ME.

Factor, Jan Robert, ed. *Biology of the Lobster Homarus americanus*. San Diego, CA: Academic Press, 1995.

Hickman, Cleveland P. *Biology of the Invertebrates*. St. Louis, MO: The C.V. Mosby Company, 1967.

Lockhart, Frank and Bruce Estrella. *Fishery Management Report No. 29 of the Atlantic States Marine Fisheries Commission*. Washington, D.C.: National Oceanic and Atmospheric Administration, 1997.

About the Author

Peter Prybot's photojournalism of fisheries has flourished, including numerous magazine cover photos and articles. Today he writes an "Ebb & Flow" column for the local newspaper, the *Gloucester Daily Times*, and is the area correspondent for the popular monthly fishing trade journal *Commercial Fisheries News*, published in Stonington, Maine. He contributes photos to *National Fisherman* and the Associated Press. His camera equipment goes with him most of the time, including while out lobstering. His first book, *White-Tipped Orange Masts*, was published in 1998 by the Curious Traveler Press in Gloucester.